Ali EL ALAMI

Electrical and electronic circuits

Ali EL ALAMI

Electrical and electronic circuits

Courses and exercises

ScienciaScripts

Publisher:
Sciencia Scripts
is a trademark of
Dodo Books Indian Ocean Ltd. and OmniScriptum S.R.L publishing group

120 High Road, East Finchley, London, N2 9ED, United Kingdom
Str. Armeneasca 28/1, office 1, Chisinau MD-2012, Republic of Moldova, Europe
Printed at: see last page
ISBN: 978-620-6-90192-1

Table of contents

Chapter I: General theorems

1. General

1.1. Definitions

1.1.1. Node

A node is a point where at least three wires meet.

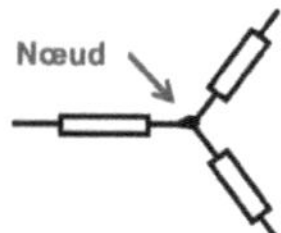

1.1.2. Branch

A branch consists of one or more dipoles connected in series between two nodes.

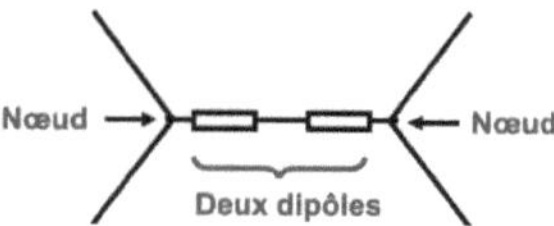

1.1.3. Mesh

A mesh is a set of branches forming a closed loop. An orientation is chosen for each mesh.

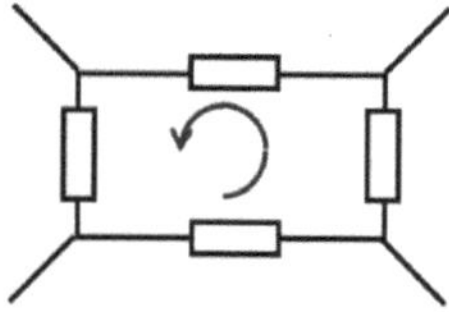

1.1.4. Network

A network, or circuit, is a set of components linked by connecting wires that can be analysed in terms of nodes, branches and meshes.

1.2. Resistance associations

1.2.1. Serial association

The equivalent resistance $R_{éq}$ of several resistors connected in series is equal to the sum of their resistances.

$$R_{éq} = \sum_{i=1}^{n} R_i$$

Important note:

In series, the equivalent resistance is greater than the greatest of the associated resistances.

1.2.2. Parallel association

The equivalent resistance $R_{éq}$ of several resistors connected in parallel is equal to the sum of their inverse resistances.

$$\frac{1}{R_{éq}} = \sum_{i=1}^{n} \frac{1}{R_i}$$

Important note:

In parallel, the equivalent resistance is smaller than the smallest of the associated resistances.

1.3. Kennelly's theorem

It establishes an equivalence between resistors placed in a delta configuration and resistors placed in a star configuration.

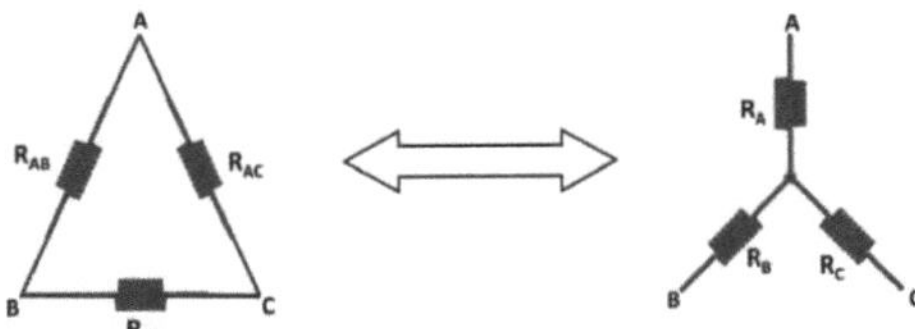

A. Triangle to star transformation

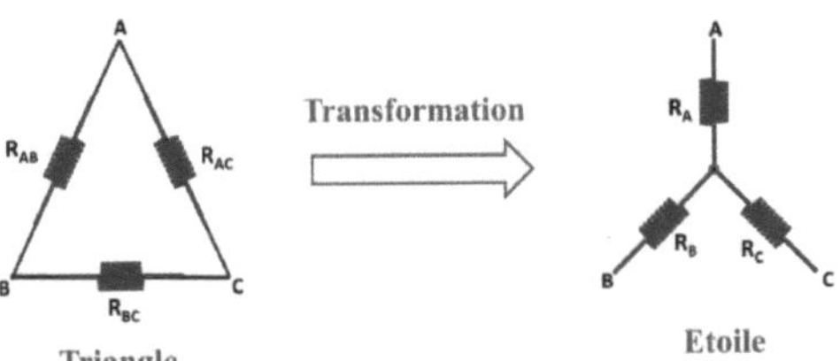

The equivalent resistance of one leg of the star is equal to the product of the adjacent resistances divided by the total sum of the resistances of the triangle.

In fact, we find the following expressions :

$$R_A = \frac{R_{AB} \cdot R_{AC}}{R_{AB} + R_{BC} + R_{AC}}$$

$$R_B = \frac{R_{AB} \cdot R_{BC}}{R_{AB} + R_{BC} + R_{AC}}$$

$$R_C = \frac{R_{BC} \cdot R_{AC}}{R_{AB} + R_{BC} + R_{AC}}$$

B. Star to triangle transformation (without demonstration)

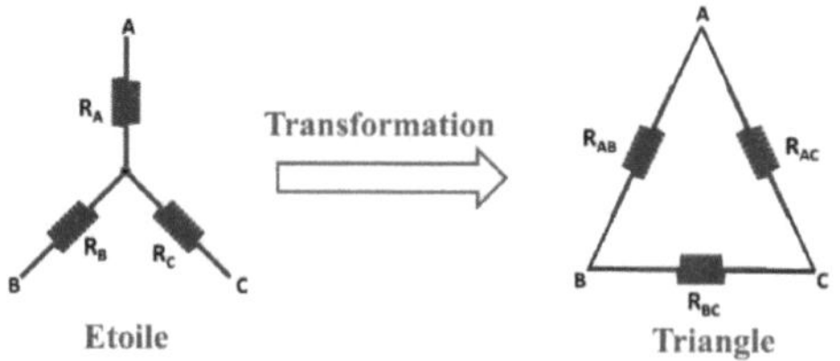

The diagram above shows the following expressions:

$$R_{AB} = \frac{R_A \cdot R_B + R_B \cdot R_C + R_C \cdot R_A}{R_C}$$

$$R_{BC} = \frac{R_A \cdot R_B + R_B \cdot R_C + R_C \cdot R_A}{R_A}$$

$$R_{AC} = \frac{R_A \cdot R_B + R_B \cdot R_C + R_C \cdot R_A}{R_B}$$

1.4. Kirchhoff's laws

In 1845, the German physicist Gustav Kirchhoff established two laws on which all calculations of electrical circuits are based: the law of nodes and the law of meshes.

1.4.1. Law of knots

The sum of the currents arriving at a node is equal to the sum of the currents leaving it.

In general, the law of knots is written as :

$$\sum_{k=1}^{n} \pm I_k = 0$$

$(+)$ if the current I_k ends at the node and $(-)$ if it leaves.

1.4.2. Law of meshes

The sum of the voltages at the terminals of the different branches of a mesh traversed in a given direction is zero.

In general, the law of meshes is written :

$$\sum_{k=1}^{n} \pm U_k = 0$$

$(+)$ if the voltage U_k is oriented in the chosen direction of the mesh and $(-)$ if the tension U_k is in the opposite direction to that chosen.

1.4.3. Application example

Consider the following electrical circuit

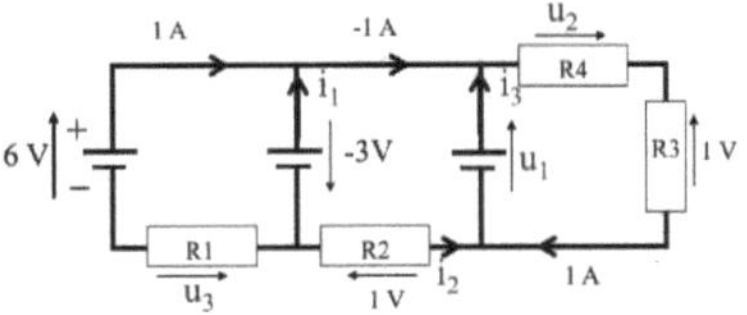

- Calculate the voltages u_1, u_2 and u_3.
- Calculate the currents i_1, i_3 and i_2.

- **Calculating voltages u_1, u_2 and u_3 :**

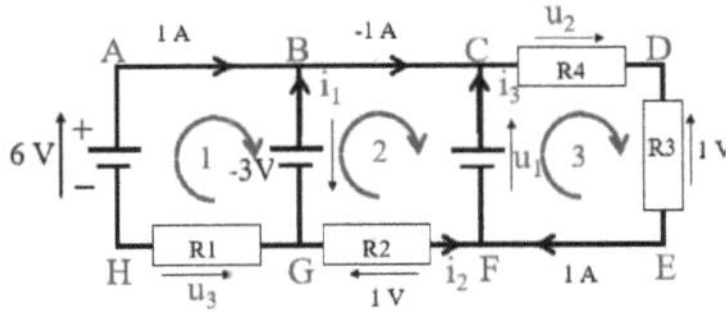

Mesh 2 [BCFG] : $-(-3) - u_1 + 1 = 0$ therefore : $u_1 = 4\ V$

Mesh 3 [CDEF] : $u_1 + u_2 - 1 = 0$ so $u_2 = -3\ V$

Mesh 1 [ABGH] : $6 + (-3) - u_3 = 0$ therefore : $u_3 = 3\ V$

- **Calculating currents i_1, i_3 and i_2 :**

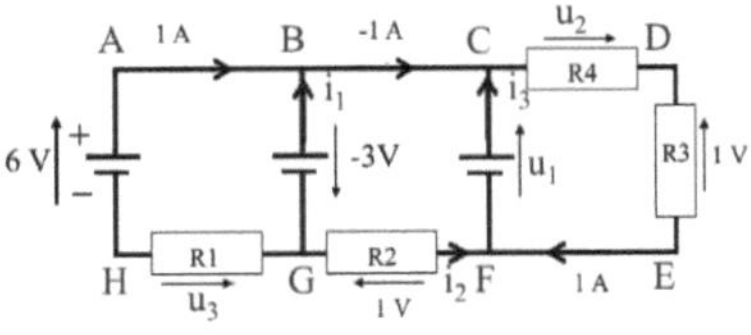

Node B : $1 + i_1 - (-1) = 0$ therefore : $i_1 = -2\ A$

Node C : $-1 - (+1) + i_3 = 0$ therefore : $i_3 = 2\ A$

Node F : $+i_2 - i_3 + 1 = 0$ therefore : $i_2 = 1\ A$

1.5. Voltage and current dividers

1.5.1. Voltage divider

The voltage divider circuit divides a voltage U into as many voltages U_i as there are resistors in series R_i.

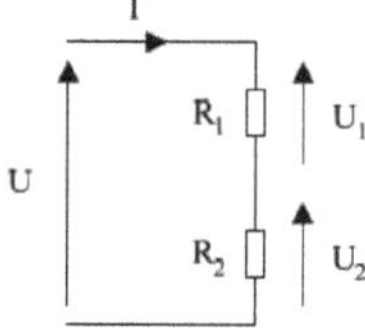

According to the circuit diagram above, we have :

$$U = U_1 + U_2 = R_1 . I + R_2 . I = (R_1 + R_2) . I$$

This gives us the following two equations:

$$U_1 = \frac{R_1}{(R_1 + R_2)} . U$$

$$U_2 = \frac{R_2}{(R_1 + R_2)} . U$$

Important notes:

- Care must be taken to apply the voltage divider formula correctly, particularly when there are parallel combinations of resistors in the circuit.
- You also need to pay attention to the direction of the voltages.

This result applies in general to the combination of n resistors connected in series.

$$U_i = \frac{R_i}{\sum_{k=1}^{n} R_k} U$$

1.5.2. Current divider

The current divider divides a current I into as many currents I_i as there are resistors in parallel R_i.

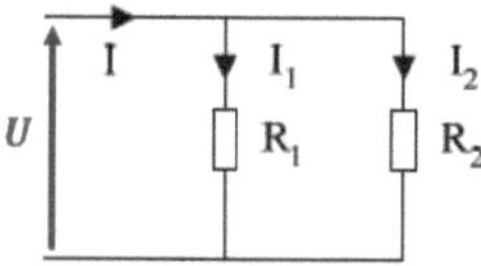

The two resistors can be combined R_1 and R_2 connected in parallel by :

$$R_{\acute{e}q} = \frac{R_1 . R_2}{R_1 + R_2}$$

The voltage across the equivalent resistor is :

$$U = R_1 . I_1 = R_2 . I_2 = R_{\acute{e}q} . I = \frac{R_1 . R_2}{(R_1 + R_2)} . I$$

This gives us the following two equations:

$$I_1 = \frac{R_2}{(R_1 + R_2)} . I$$

$$I_2 = \frac{R_1}{(R_1 + R_2)} . I$$

Important notes:

- Care must be taken to apply the current divider formula correctly, particularly when there are combinations of resistors in series in the circuit.
- Attention must also be paid to the orientation of the intensities.

This result applies in general to the combination of n resistors connected in parallel.

$$I_i = \frac{G_i}{\sum_{k=1}^{n} G_k} I$$

2. General theorems

2.1. Thevenin's theorem

2.1.1. Theorem statement

Any linear circuit can be modelled by a voltage source in series with a resistor.

2.1.2. Principle of Thévenin's theorem

This theorem is useful when defining the current I and/or voltage U in a branch of an electric circuit.

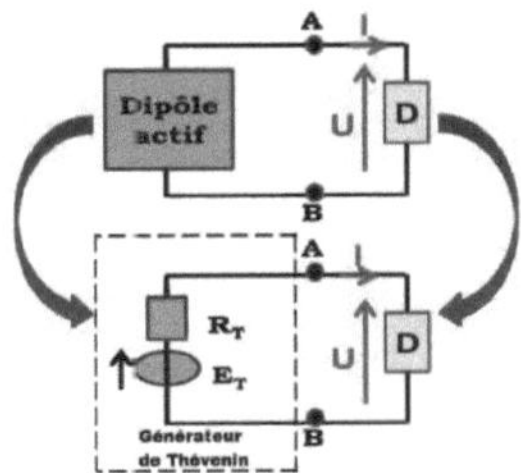

With :

E_T Potential difference across the dipole when it is at no-load.

R_T Equivalent resistance as seen from the terminals of the dipole when all its sources are cancelled (switched off).

2.1.3. Examples of applications

A. Application example 1

Consider the following electrical circuit:

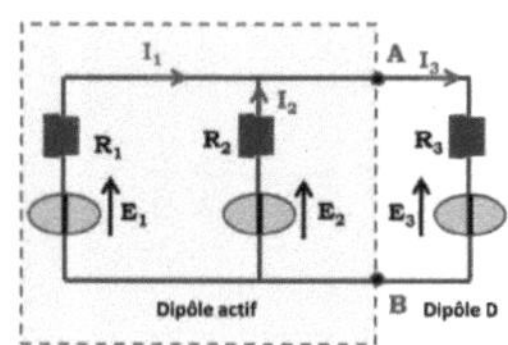

Determine the expression for the current I_3 by applying Thévenin's theorem.

❖ **Determination of E_T :**

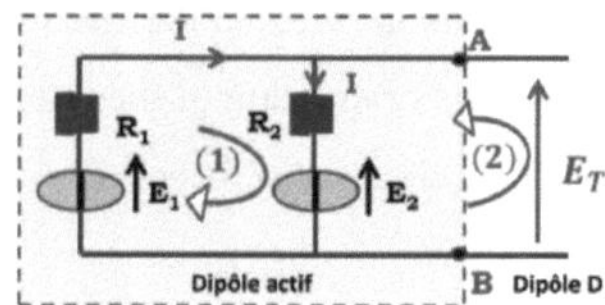

By applying the law of meshes, we find :

- Maille (1) : $I = \dfrac{E_1 - E_2}{R_1 + R_2}$

- Maille (2) : $E_T = E_2 + R_2.I$

Hence :

$$E_T = \frac{R_1 . E_2 + R_2 . E_1}{R_1 + R_2}$$

❖ **Determination of R_T :**

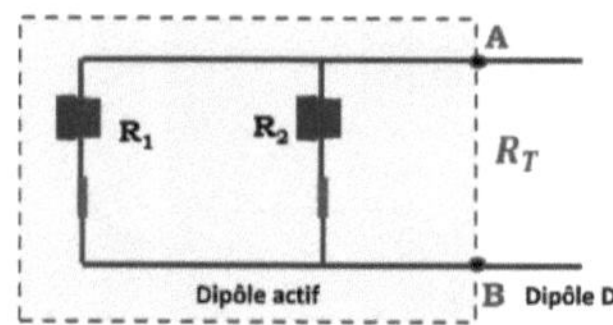

According to the circuit diagram above, we have :

$$R_T = R_1 \; // \; R_2$$

So :

$$R_T = \frac{R_1.R_2}{R_1 + R_2}$$

❖ **Determination of I_3 :**

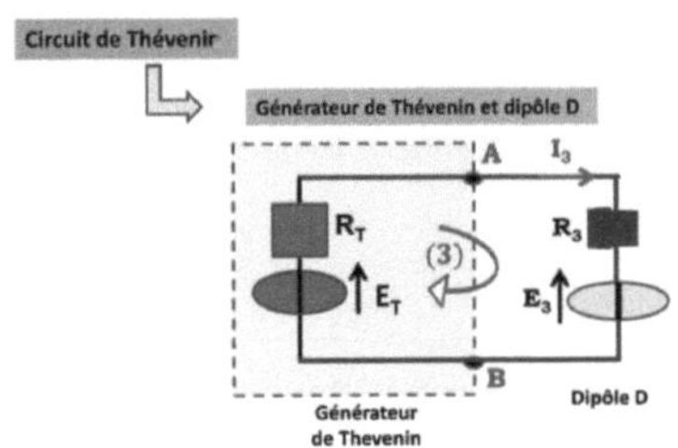

Mesh (3) gives us :

$$E_T - R_T.I_3 - R_3.I_3 - E_3 = 0$$

Hence :

$$I_3 = \frac{E_T - E_3}{R_T + R_3}$$

B. Application example 2

Consider the following electrical circuit:

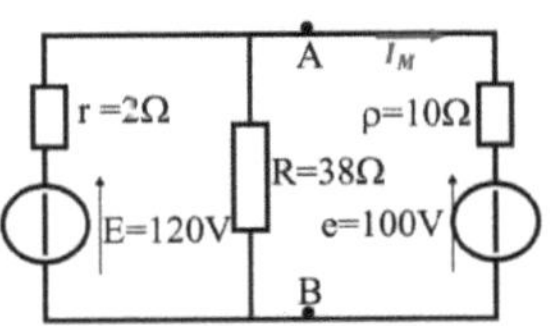

Calculate the current I_M by applying Thévenin's theorem.

Similarly, we find :

$$R_T = \frac{r.R}{r + R}$$

A.N. : $R_T = 1.9\ \Omega$

$$E_T = \frac{R}{(r + R)}.E$$

A.N. : $E_T = 114\ V$

Consequently, the Thévenin circuit becomes :

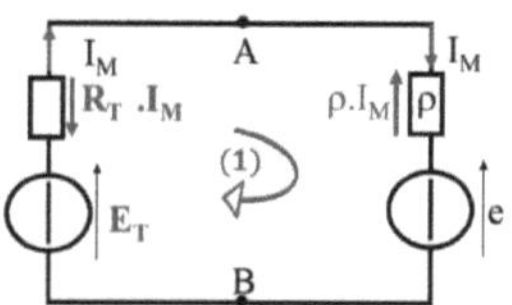

Mesh (1) gives us :

$$E_T - R_T.I_M - \rho.I_M - e = 0$$

This gives us :

$$I_M = \frac{E_T - e}{R_T + \rho}$$

Hence :

$$I_M = 1.17\ A$$

2.2. Norton's theorem

2.2.1. Theorem statement

Any linear circuit can be modelled by a current source in parallel with a resistor.

2.2.2. Norton theorem principle

This theorem is useful when defining the current I and/or voltage U in a branch of an electric circuit.

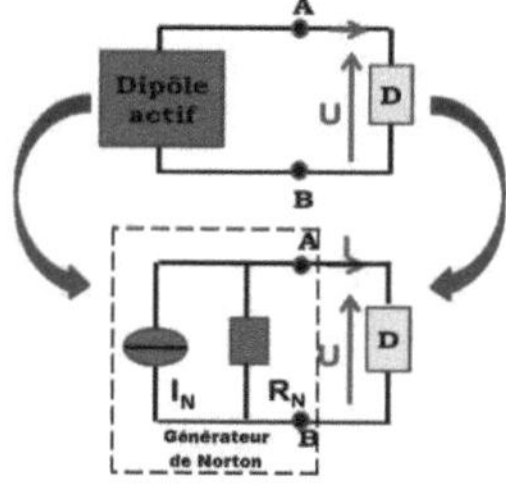

With :

I_N Current flowing through the dipole when it is short-circuited.

R_N Equivalent resistance as seen from the terminals of the dipole when all its sources are switched off.

2.2.3. Application example

Consider the following electrical circuit:

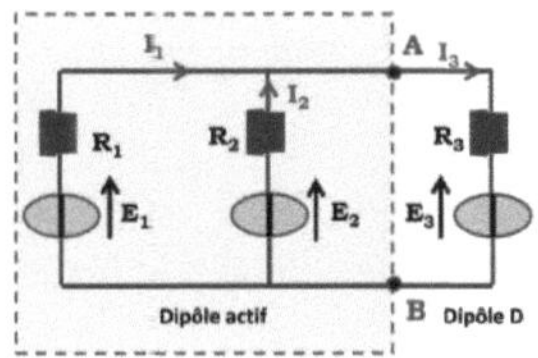

Determine the expression for the current I_3 by applying Norton's theorem.

❖ ***Determination of I_N :***

Remove the third leg and short-circuit A and B, then calculate the current I_N.

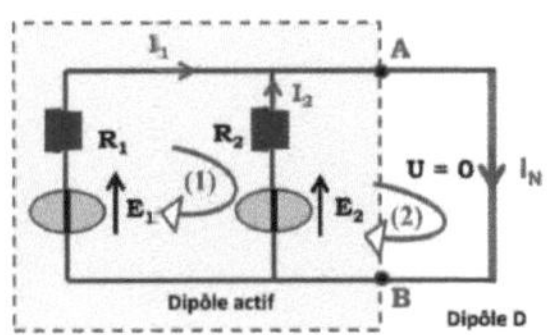

According to the law of knots, we have :

$$I_N = I_1 + I_2$$

By applying the law of meshes, we find :

- Maille (1) : $I_1 = \dfrac{E_1}{R_1}$

- Maille (2) : $I_2 = \dfrac{E_2}{R_2}$

Hence :

$$I_N = \frac{E_1}{R_1} + \frac{E_2}{R_2}$$

❖ ***Determination of R_N :***

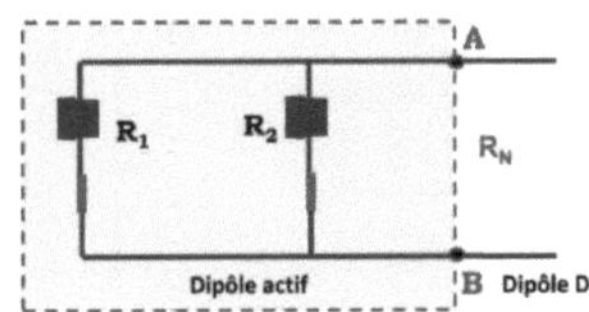

According to the circuit diagram above, we have :

$$R_N = R_1 \mathbin{/\!/} R_2$$

So :

$$R_N = \frac{R_1.R_2}{R_1 + R_2}$$

❖ *Determination of I_3 :*

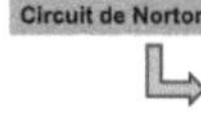
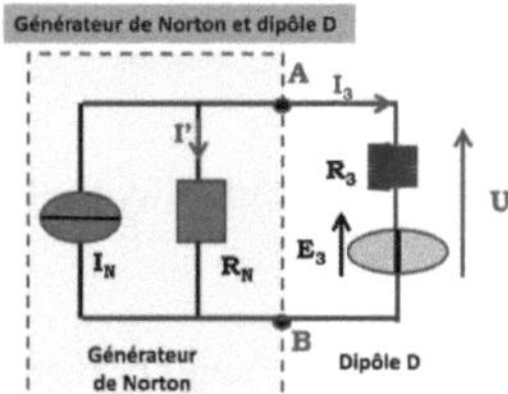

According to Norton's circuit, we have :

$$I_3 = I_N - I'$$

Or :

$$I' = \frac{U}{R_N} = \frac{E_3 + R_3.I_3}{R_N}$$

This gives us :

$$I_3 = I_N - \frac{(E_3 + R_3.I_3)}{R_N}$$

Hence :

$$I_3 = \frac{I_N - \dfrac{E_3}{R_N}}{1 + \dfrac{R_3}{R_N}}$$

2.2.4. Equivalence between Thévenin and Norton generators

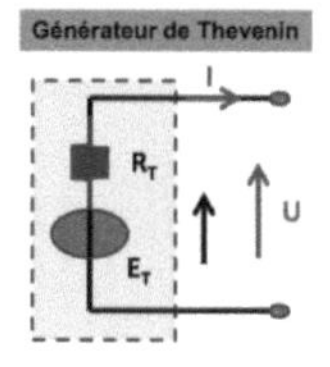

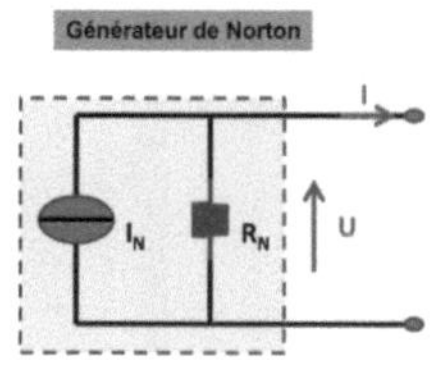

Thévenin generator	Norton generator
$U = E_T - R_T.I$	$U = R_N.I_N - R_N.I$
$I = \dfrac{E_T}{R_T} - \dfrac{U}{R_T}$	$I = I_N - \dfrac{U}{R_N}$

$E_T = R_N \cdot I_N$ $R_T = R_N$	$I_N = \dfrac{E_T}{R_T}$ $R_N = R_T$

2.3. Superposition theorem

2.3.1. Theorem statement

The voltage (or current) between two points in a linear electrical circuit comprising several sources is equal to the algebraic sum of the voltages (or currents) obtained between these two points when each source acts alone.

2.3.2. Method of extinguishing voltage and current sources

A. Theoretical extinction of a voltage source

Ideal voltage generators are replaced by short circuits.

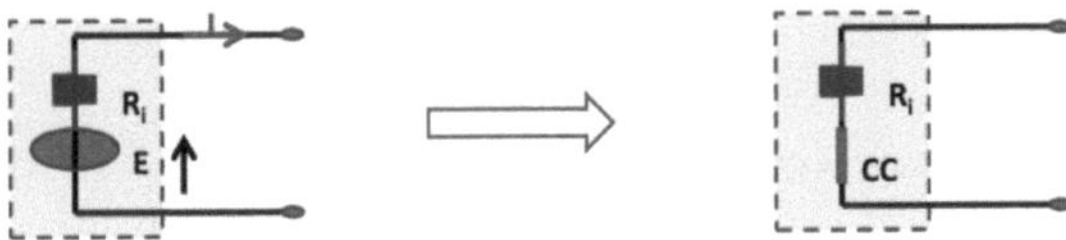

B. Theoretical extinction of a current source

Ideal current generators are replaced by open circuits.

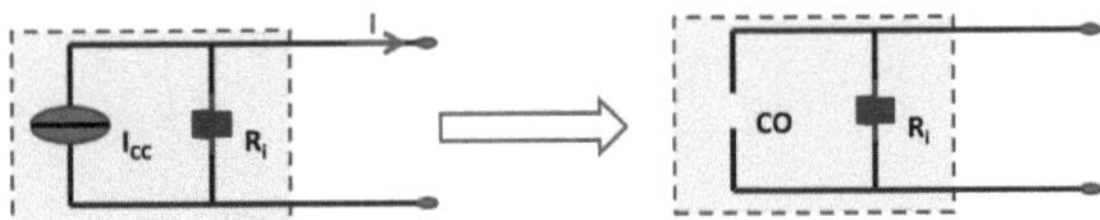

2.3.3. Examples of applications

A. Application example 1

Consider the following electrical circuit:

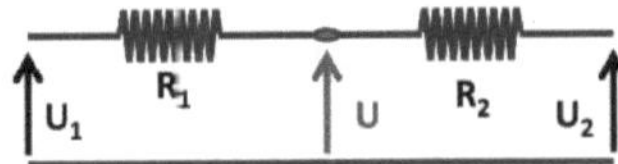

Find the expression for the voltage U by applying the superposition theorem.

Switching off the voltage source U_2 :

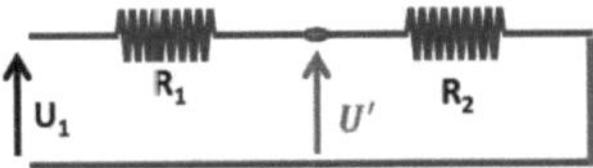

We apply the voltage divider rule and find :

$$U' = \frac{R_2 . U_1}{R_1 + R_2}$$

Switching off the voltage source U_1 :

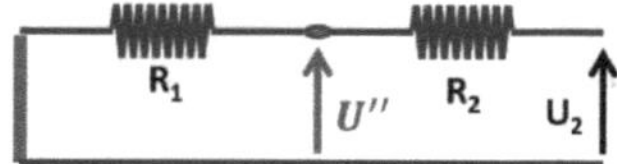

Similarly, there are :

$$U'' = \frac{R_1 . U_2}{R_1 + R_2}$$

Finally:

$$U = U' + U'' = \frac{R_2 . U_1 + R_1 . U_2}{R_1 + R_2}$$

B. Application example 2

Consider the following electrical circuit:

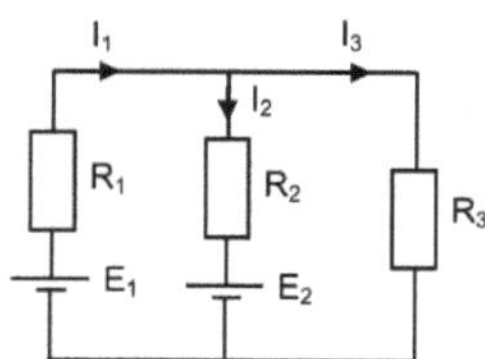

Data : $R_1 = 2\ \Omega$, $R_2 = 5\ \Omega$, $R_3 = 10\ \Omega$, $E_1 = 20\ V$ and $E_2 = 70\ V$.

Calculate the currents I_1, I_2 and I_3 by applying the superposition theorem.

According to the principle of the superposition theorem, the original circuit will be divided into the following two sub-circuits:

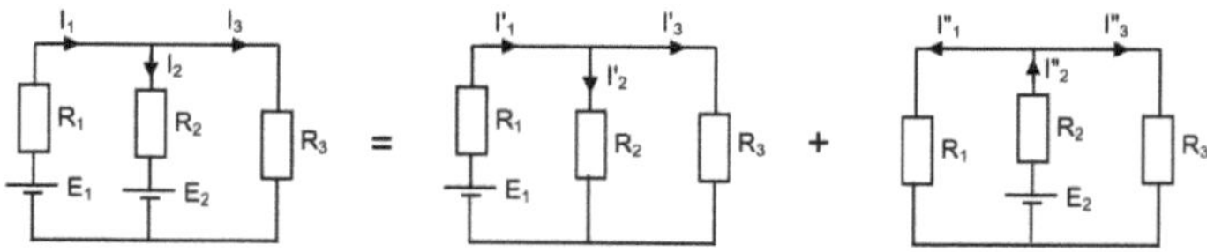

Real currents I_1, I_2 and I_3 are given by :

$$\begin{cases} I_1 = I_1' - I_1'' \\ I_2 = I_2' - I_2'' \\ I_3 = I_3' + I_3'' \end{cases}$$

14

With :

$$\begin{cases} I'_1 = \dfrac{E_1}{R_1+\left(\frac{R_2.R_3}{R_2+R_3}\right)} = 3.75\ A \\[3mm] I'_2 = \dfrac{R_3}{(R_2+R_3)}.I'_1 = 2.5\ A \\[3mm] I'_3 = \dfrac{R_2}{(R_2+R_3)}.I'_1 = 1.25\ A \end{cases} \quad \text{and} \quad \begin{cases} I''_2 = \dfrac{E_2}{R_2+\left(\frac{R_1.R_3}{R_1+R_3}\right)} = 10.5\ A \\[3mm] I''_1 = \dfrac{R_3}{(R_1+R_3)}.I''_2 = 8.75\ A \\[3mm] I''_3 = \dfrac{R_1}{(R_1+R_3)}.I''_2 = 1.75\ A \end{cases}$$

Hence :

$$\begin{cases} I_1 = -5\ A \\ I_2 = -8\ A \\ I_3 = 3\ A \end{cases}$$

Important notes:

- The current I_1 is negative. So its true direction is the opposite of the chosen direction.
- The current I_2 is negative. So its true direction is the opposite of the chosen direction.

2.4. Millman's theorem

2.4.1. Theorem statement

Millman's theorem applies to an electrical circuit consisting of n branches in parallel. Each of these branches comprises a perfect voltage generator in series with a linear element (e.g. a resistor).

2.4.2. Demonstration

Consider the following circuit:

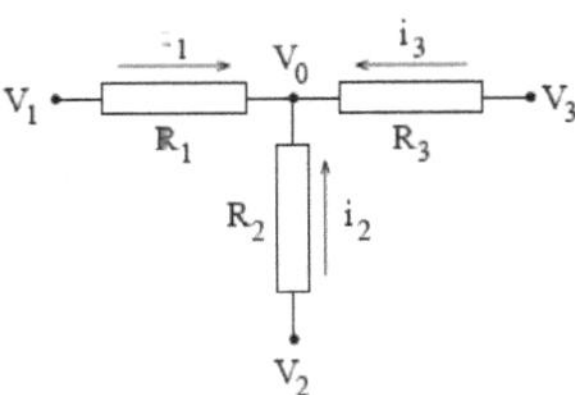

For each branch, we can write :

$$\begin{cases} V_1 - V_0 = R_1.i_1 \\ V_2 - V_0 = R_2.i_2 \\ V_3 - V_0 = R_3.i_3 \end{cases}$$

Or :

$$\begin{cases} i_1 = \dfrac{V_1 - V_0}{R_1} \\[2mm] i_2 = \dfrac{V_2 - V_0}{R_2} \\[2mm] i_3 = \dfrac{V_3 - V_0}{R_3} \end{cases}$$

Since :

$$i_1 + i_2 + i_3 = 0$$

So :

$$\frac{V_1 - V_0}{R_1} + \frac{V_2 - V_0}{R_2} + \frac{V_3 - V_0}{R_3} = 0$$

Hence :

$$V_0 = \frac{\dfrac{V_1}{R_1} + \dfrac{V_2}{R_2} + \dfrac{V_3}{R_3}}{\dfrac{1}{R_1} + \dfrac{1}{R_2} + \dfrac{1}{R_3}}$$

In general, the potential V_0 is expressed as a function of the potentials at neighbouring nodes as follows:

$$V_0 = \frac{\dfrac{V_1}{R_1} + \dfrac{V_2}{R_2} + \cdots + \dfrac{V_n}{R_n}}{\dfrac{1}{R_1} + \dfrac{1}{R_2} + \cdots + \dfrac{1}{R_n}} = \frac{\sum_{i=1}^{n} \dfrac{V_i}{R_i}}{\sum_{i=1}^{n} \dfrac{1}{R_i}}$$

The conductance of a resistive dipole can also be defined by the inverse of its resistance. Let :

$$G_i = \frac{1}{R_i} \quad \text{(Unit: Siemens (S))}$$

Millman's theorem can also be written as :

$$V_0 = \frac{\sum_{i=1}^{n} G_i . V_i}{\sum_{i=1}^{n} G_i}$$

2.4.3. Application example

Consider the following electrical circuit:

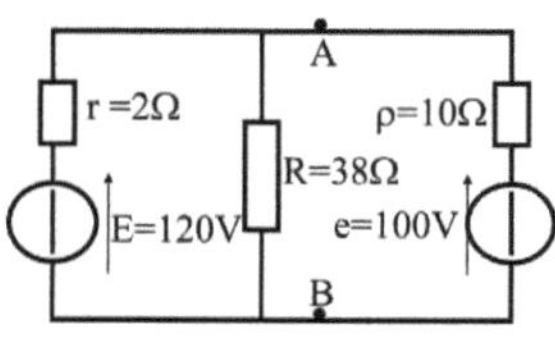

Calculate the value of the voltage U_{AB} by applying Millman's theorem.

According to Millman's theorem, we find :

$$U_{AB} = \frac{\dfrac{E}{r} + \dfrac{0}{R} + \dfrac{e}{\rho}}{\dfrac{1}{r} + \dfrac{1}{R} + \dfrac{1}{\rho}}$$

A.N. :

$$U_{AB} = \frac{\dfrac{120}{2} + \dfrac{0}{38} + \dfrac{100}{10}}{\dfrac{1}{2} + \dfrac{1}{38} + \dfrac{1}{10}}$$

Hence :

$$U_{AB} = 111.76\,V$$

Chapter II: Alternating sinusoidal system

1. Characteristics of periodic and sinusoidal signals

1.1. Characteristic of a periodic signal

1.1.1. Frequency and pulse

Consider the following periodic signal $u(t)$ signal :

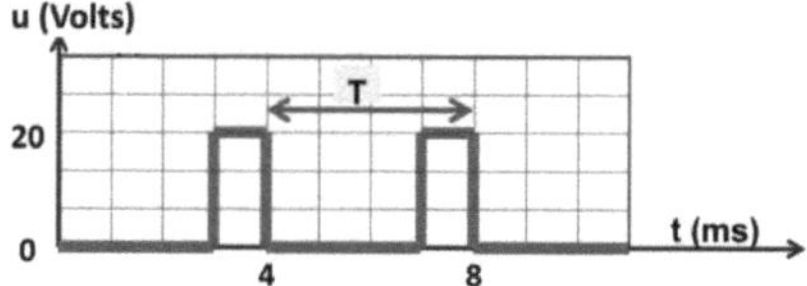

The frequency f (in hertz) corresponds to the number of periods per unit of time:

$$f = \frac{1}{T}$$

The pulsation ω (in radians per second) is defined by :

$$\omega = 2\pi f = \frac{2\pi}{T}$$

1.1.2. Average value

For a periodic signal $u(t)$ of period T. The mean value is defined by :

$$U_{moy} = \frac{1}{T} \int_0^T u(t).dt$$

Application example:

Consider the following periodic signal $u(t)$ signal:

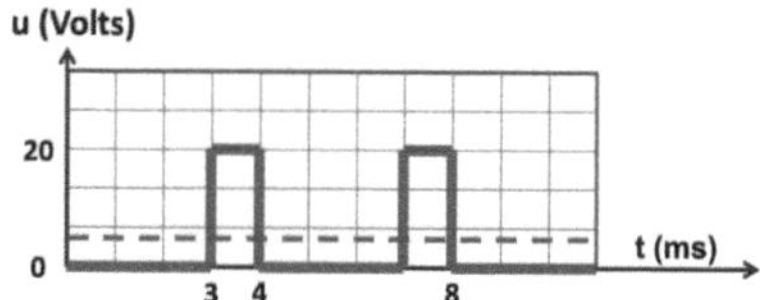

Calculate the mean value of the signal $u(t)$.

By definition, we have :

$$U_{moy} = \frac{1}{T} \int_0^T u(t).dt$$

So :

$$U_{moy} = \frac{1}{4} \int_3^4 20.dt = \frac{20}{4} = 5\ V$$

1.1.3. RMS value

For a periodic signal $u(t)$ of period T. The rms value is defined by :

$$U_{eff} = \sqrt{\frac{1}{T} \int_0^T u^2(t).dt}$$

Application example:

Consider the following periodic signal $u^2(t)$ signal:

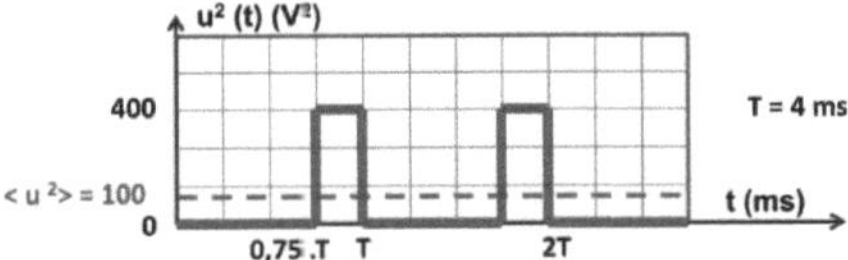

Calculate the rms value of this signal.

By definition, we have :

$$U_{eff} = \sqrt{\frac{1}{T} \int_0^T u^2(t).dt}$$

So :

$$U_{eff} = \sqrt{\frac{1}{T} \int_{0.75T}^T 400.dt} = 10\ V$$

1.2. Characteristics of a sinusoidal signal

Consider the following sinusoidal signal $u(t) = U_m \cos(\omega t + \varphi)$ signal:

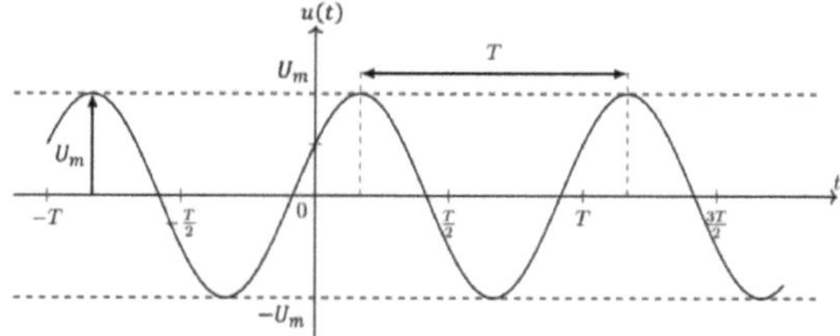

With :

$\omega t + \varphi$: Signal phase (angle in radians)

φ Phase at $t = 0$ (in radians)

ω Signal pulse (in radians per second)

T Signal period (in seconds)

f Signal frequency (in Hertz)

1.2.1. Average value

For a sinusoidal signal $u(t)$ of period T. The mean value is defined by :

$$U_{moy} = \frac{1}{T} \int_0^T u(t).dt$$

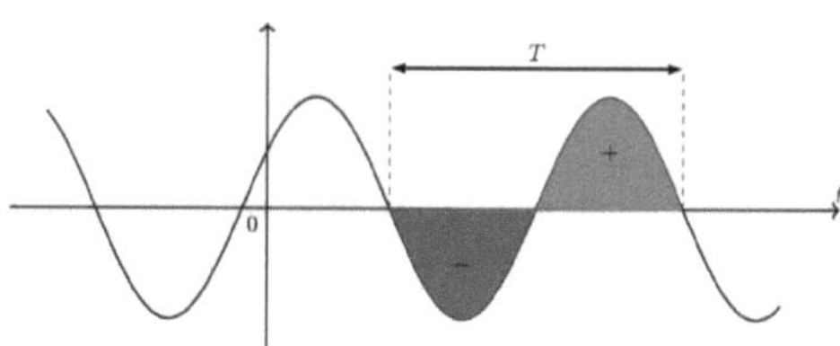

Over a period T the area under the curve is zero (the positive area exactly offsetting the negative area).

The mean value of a sine or cosine is zero.

1.2.2. RMS value

For a sinusoidal signal $u(t)$ of period T. The rms value is defined by :

$$U_{eff} = \sqrt{\frac{1}{T} \int_0^T u^2(t).dt}$$

In our case, we have :

$$u(t) = U_m \cos(\omega t + \varphi)$$

$$U_{eff} = \sqrt{\frac{U_m^2}{T} \int_0^T \left(\cos^2(\omega t + \varphi)\right).dt}$$

$$U_{eff} = \sqrt{\frac{U_m^2}{T} \int_0^T \left(\frac{1 + \cos(2\omega t)}{2}\right).dt}$$

After simplifying the calculation, we obtain :

$$U_{eff} = \frac{U_m}{\sqrt{2}}$$

Application example:

Consider the following sinusoidal signal $u(t)$ signal:

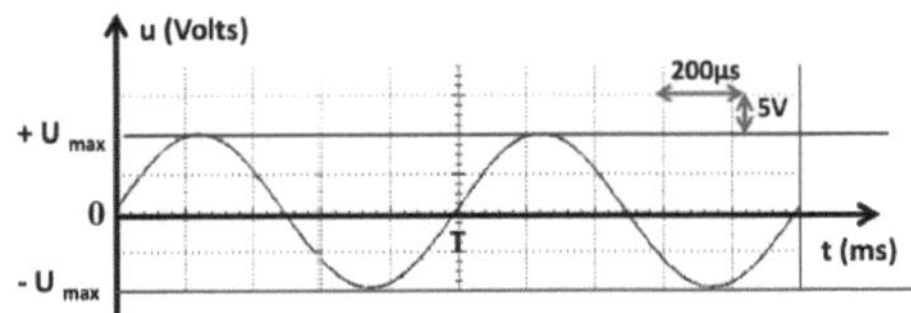

Calculate the rms value of this signal.

We know that :

$$U_{eff} = \frac{U_m}{\sqrt{2}}$$

With : $U_m = 10\ V$

Hence :

$$U_{eff} = 7.07\ V$$

2. Complex impedance of a linear dipole under sinusoidal conditions

Consider the following linear dipole in sinusoidal operation:

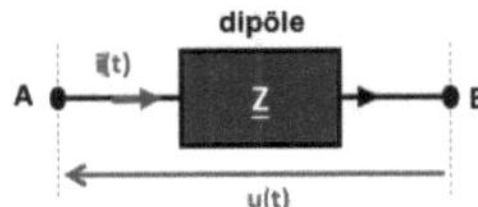

With :

$$u(t) = U_m \cos(\omega t + \varphi) \quad \text{and} \quad i(t) = I_m \cos(\omega t + \varphi')$$

Under sinusoidal conditions, the complex impedance $\underline{Z}$ of a linear dipole is defined by :

$$\underline{Z} = \frac{\underline{u}}{\underline{i}}$$

The module of $\underline{Z}$ is :

$$Z = \frac{U_m}{I_m}$$

The argument of $\underline{Z}$ is :

$$arg(\underline{Z}) = \varphi_{u/i}$$

With : $\varphi_{u/i} = \varphi - \varphi'$

3. Complex admittance of a linear dipole under sinusoidal conditions

Consider the following linear dipole in sinusoidal mode:

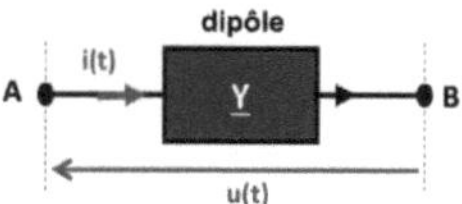

With :

$$u(t) = U_m \cos(\omega t + \varphi)$$

and

$$i(t) = I_m \cos(\omega t + \varphi')$$

In sinusoidal operation, the complex admittance $\underline{Y}$ of a linear dipole is defined by :

$$\underline{Y} = \frac{\underline{i}}{\underline{u}}$$

The module of $\underline{Y}$ is :

$$Y = \frac{I_m}{U_m}$$

The argument of $\underline{Y}$ is :

$$arg(\underline{Y}) = \varphi_{i/u}$$

With : $\varphi_{i/u} = \varphi' - \varphi$

4. Examples of elementary dipoles in sinusoidal operation

Elementary dipole	Complex impedance
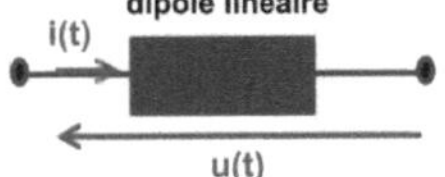	$\underline{Z}_R = R$
	$\underline{Z}_L = jL\omega$
	$\underline{Z}_C = \dfrac{1}{jC\omega}$

5. Electrical power in sinusoidal mode

5.1. Instantaneous power

The instantaneous power received by a linear dipole at an instant t is defined by :

$$p(t) = u(t).i(t)$$

$$p(t) = \frac{U_m \cdot I_m}{2} \cos(2\omega t + \varphi + \varphi') + \frac{U_m \cdot I_m}{2} \cos(\varphi - \varphi')$$

With :

$$u(t) = U_m \cos(\omega t + \varphi) \quad \text{and} \quad i(t) = I_m \cos(\omega t + \varphi')$$

5.2. Average power

Average power P_{moy} is defined by :

$$P_{moy} = \frac{1}{T} \int_0^T p(t).dt = \frac{1}{T} \int_0^T (u(t).i(t)).dt$$

We replace $p(t)$ by its expression and we obtain :

$$P_{moy} = \frac{1}{T} \left[\int_0^T \left(\frac{U_m \cdot I_m}{2} \cos(2\omega t + \varphi + \varphi') + \frac{U_m \cdot I_m}{2} \cos(\varphi - \varphi') \right).dt \right]$$

The integral over a period T of $\cos(2\omega t + \varphi + \varphi')$ is zero.

So :

$$P_{moy} = \frac{1}{T} \int_{0}^{T} \left(\frac{U_m \cdot I_m}{2} \cos(\varphi_{u/i}) \right) . dt$$

With : $\varphi_{u/i} = \varphi - \varphi'$

Hence :

$$P_{moy} = \frac{U_m \cdot I_m}{2} \cos(\varphi_{u/i})$$

The term $\cos(\varphi_{u/i})$ is called the power factor.

Chapter III: Electric quadrupoles

1. Definition and representation

An electrical quadripole is a complex electrical circuit with two input terminals and two output terminals.

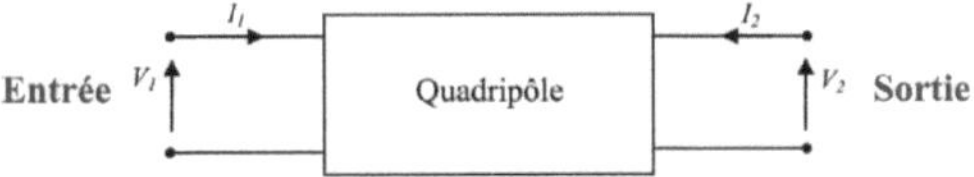

The electrical quantities of the quadrupole are :

- The input I_1 and output currents I_2 and the input and output voltages V_1 and output voltages V_2.
- By convention, the currents entering the quadripole are in the positive direction.

2. Examples of quadrupoles

2.1. Serial quadrupole

Consider the following series quadrupole:

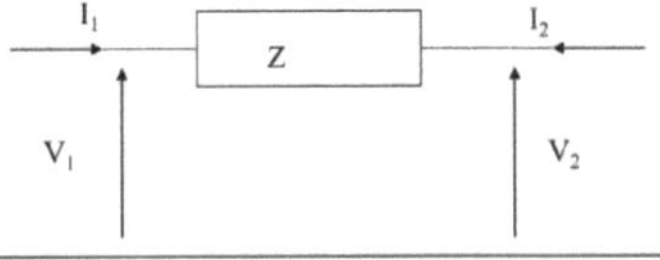

It contains a single impedance. The law of meshes gives :

$$V_2 = V_1 - Z.I_1 \quad \text{and} \quad I_2 = -I_1$$

Or in matrix form :

$$\begin{cases} V_2 = V_1 - Z.I_1 \\ I_2 = 0.V_1 - I_1 \end{cases}$$

$$\begin{bmatrix} V_2 \\ I_2 \end{bmatrix} = \begin{bmatrix} 1 & Z \\ 0 & 1 \end{bmatrix} \begin{bmatrix} V_1 \\ -I_1 \end{bmatrix}$$

Or :

$$\begin{bmatrix} V_2 \\ I_2 \end{bmatrix} = [T] \begin{bmatrix} V_1 \\ -I_1 \end{bmatrix}$$

The determinant of the matrix $[T]$ linking the input quantities to the output quantities is equal to 1.

2.2. Parallel quadrupole

Consider the following parallel quadrupole:

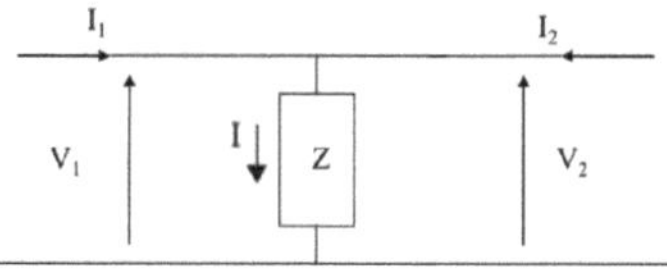

The law of meshes gives :

$$V_2 = V_1 = Z.I$$

So :

$$I = \frac{V_1}{Z}$$

The law of knots gives :

$$I = I_1 + I_2$$

Hence :

$$I_2 = \frac{V_1}{Z} - I_1$$

Or in matrix form :

$$\begin{cases} V_2 = V_1 - 0.I_1 \\ I_2 = \frac{1}{Z}.V_1 - I_1 \end{cases}$$

$$\begin{bmatrix} V_2 \\ I_2 \end{bmatrix} = \begin{bmatrix} 1 & 0 \\ 1/Z & 1 \end{bmatrix} \begin{bmatrix} V_1 \\ -I_1 \end{bmatrix}$$

Or :

$$\begin{bmatrix} V_2 \\ I_2 \end{bmatrix} = [T] \begin{bmatrix} V_1 \\ -I_1 \end{bmatrix}$$

The determinant of the matrix $[T]$ is 1.

3. Representative matrices for quadrupoles

3.1. Impedance matrix [Z] and equivalent electrical diagram

The input V_1 and output voltages V_2 are expressed as a function of the input and output currents I_1 and output currents I_2 via the impedance parameters.

The characteristic equations of this quadrupole can be expressed as :

$$V_1 = Z_{11}.I_1 + Z_{12}.I_2$$
$$V_2 = Z_{21}.I_1 + Z_{22}.I_2$$

Or in matrix form :

$$\begin{bmatrix} V_1 \\ V_2 \end{bmatrix} = \begin{bmatrix} Z_{11} & Z_{12} \\ Z_{21} & Z_{22} \end{bmatrix} \begin{bmatrix} I_1 \\ I_2 \end{bmatrix} = [Z] \begin{bmatrix} I_1 \\ I_2 \end{bmatrix}$$

The equivalent electrical diagram is :

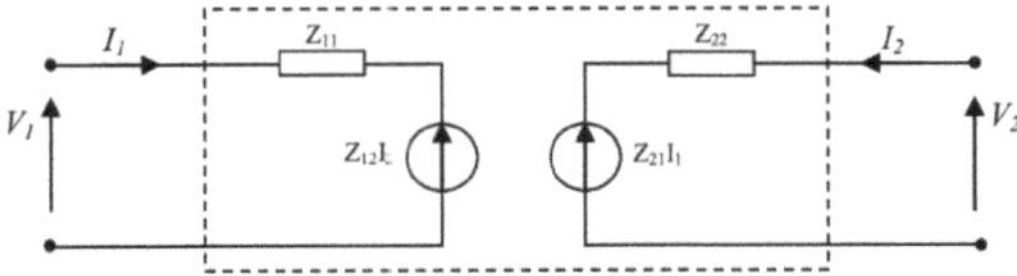

The four impedance parameters Z_{ij} parameters are defined as follows:

$Z_{11} = \dfrac{V_1}{I_1}\bigg| I_2 = 0$ Open circuit input impedance (no-load output)

$Z_{12} = \dfrac{V_1}{I_2}\bigg| I_1 = 0$ Open circuit reverse transfer impedance

$Z_{21} = \dfrac{V_2}{I_1}\bigg| I_2 = 0$ Direct open-circuit transfer impedance

$Z_{22} = \dfrac{V_2}{I_2}\bigg| I_1 = 0$ Open circuit output impedance

3.2. Admittance matrix [Y] and equivalent electrical diagram

The input I_1 and output currents I_2 are expressed as a function of the input and output voltages V_1 and output voltages V_2 via the admittance parameters.

The characteristic equations of this quadrupole can be expressed as :

$$I_1 = Y_{11}.V_1 + Y_{12}.V_2$$
$$I_2 = Y_{21}.V_1 + Y_{22}.V_2$$

Or in matrix form :

$$\begin{bmatrix} I_1 \\ I_2 \end{bmatrix} = \begin{bmatrix} Y_{11} & Y_{12} \\ Y_{21} & Y_{22} \end{bmatrix} \begin{bmatrix} V_1 \\ V_2 \end{bmatrix} = [Y] \begin{bmatrix} V_1 \\ V_2 \end{bmatrix}$$

The equivalent electrical diagram is :

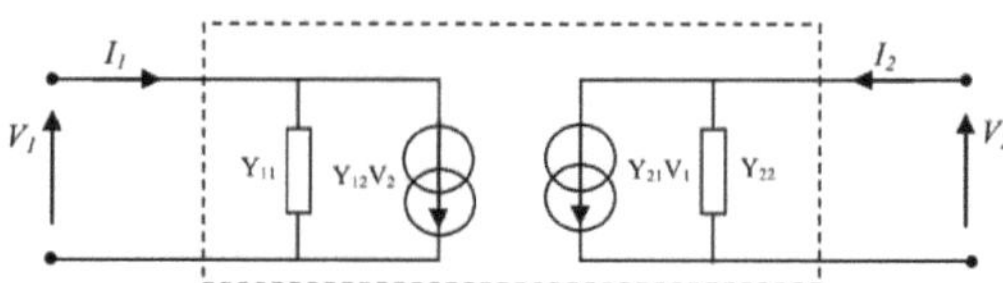

The four admittance parameters Y_{ij} are defined as follows:

$Y_{11} = \dfrac{I_1}{V_1}\bigg| V_2 = 0$ Input admittance

$Y_{12} = \dfrac{I_1}{V_2}\bigg| V_1 = 0$ Inverse transfer admittance

$Y_{21} = \dfrac{I_2}{V_1}\bigg| V_2 = 0$ Direct transfer admittance

$Y_{22} = \dfrac{I_2}{V_2}\bigg| V_1 = 0$ Output admittance

3.3. Hybrid matrix [h] and equivalent electrical diagram

The input voltage V_1 and the output current I_2 are expressed as a function of the output voltage V_2 and the input current I_1 using hybrid parameters.

The characteristic equations of this quadrupole can be expressed as :

$$V_1 = h_{11}.I_1 + h_{12}.V_2$$
$$I_2 = h_{21}.I_1 + h_{22}.V_2$$

Or in matrix form :

$$\begin{bmatrix} V_1 \\ I_2 \end{bmatrix} = \begin{bmatrix} h_{11} & h_{12} \\ h_{21} & h_{22} \end{bmatrix} \begin{bmatrix} I_1 \\ V_2 \end{bmatrix} = [h] \begin{bmatrix} V_1 \\ V_2 \end{bmatrix}$$

The equivalent electrical diagram is :

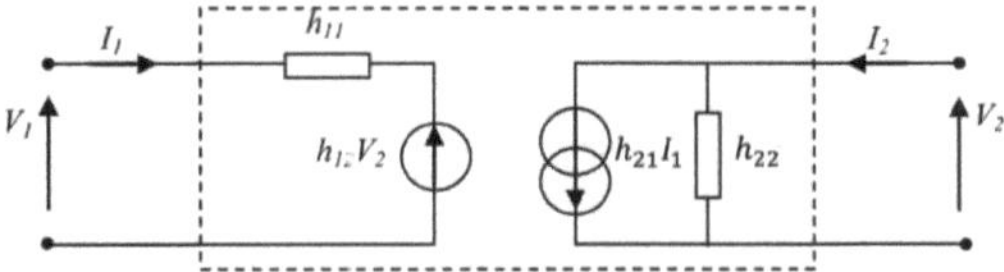

The four hybrid parameters h_{ij} are defined as follows:

$h_{11} = \left.\frac{V_1}{I_1}\right| V_2 = 0$ Input impedance

$h_{12} = \left.\frac{V_1}{V_2}\right| I_1 = 0$ Inverse voltage gain

$h_{21} = \left.\frac{I_2}{I_1}\right| V_2 = 0$: Current gain

$h_{22} = \left.\frac{I_2}{V_2}\right| I_1 = 0$ Output admittance

3.4. Transfer matrix [T]

The output values V_2 and I_2 are expressed as a function of the input variables V_1 and I_1.

The characteristic equations of this quadrupole can be expressed as :

$$V_2 = T_{11}.V_1 - T_{12}.I_1$$
$$I_2 = T_{21}.V_1 - T_{22}.I_1$$

Or in matrix form :

$$\begin{bmatrix} V_2 \\ I_2 \end{bmatrix} = \begin{bmatrix} T_{11} & T_{12} \\ T_{21} & T_{22} \end{bmatrix} \begin{bmatrix} V_1 \\ -I_1 \end{bmatrix} = [T] \begin{bmatrix} V_1 \\ -I_1 \end{bmatrix}$$

The four parameters T_{ij} are defined as follows:

$T_{11} = \left.\frac{V_2}{V_1}\right| I_1 = 0$ Voltage gain

$T_{12} = -\left.\frac{V_2}{I_1}\right| V_1 = 0$ Impedance

$T_{21} = \left.\frac{I_2}{V_1}\right| I_1 = 0$ Admittance

$T_{22} = -\left.\frac{I_2}{I_1}\right| V_1 = 0$: Currentgain

4. Association of quadripoles in cascade

The two outputs of the first quadrupole are connected to the two inputs of the second. The transfer matrices $[T_1]$ and $[T_2]$ of the two associated quadripoles.

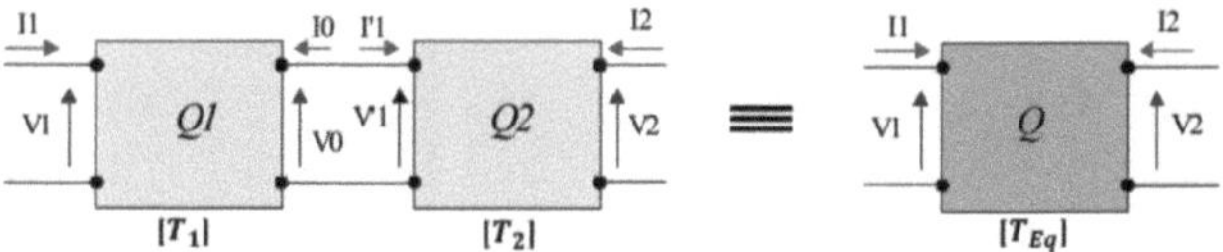

The transfer matrix of the first quadrupole :

$$\begin{bmatrix} V_0 \\ I_0 \end{bmatrix} = \begin{bmatrix} T_{11}^{(1)} & T_{12}^{(1)} \\ T_{21}^{(1)} & T_{22}^{(1)} \end{bmatrix} \begin{bmatrix} V_1 \\ -I_1 \end{bmatrix} = [T_1]\begin{bmatrix} V_1 \\ -I_1 \end{bmatrix}$$

The transfer matrix of the second quadrupole :

$$\begin{bmatrix} V_2 \\ I_2 \end{bmatrix} = \begin{bmatrix} T_{11}^{(2)} & T_{12}^{(2)} \\ T_{21}^{(2)} & T_{22}^{(2)} \end{bmatrix} \begin{bmatrix} V_1' \\ -I_1' \end{bmatrix} = [T_2]\begin{bmatrix} V_1' \\ -I_1' \end{bmatrix}$$

Or, $V_0 = V_1'$ and $I_0 = -I_1'$

We replace V_1' and $-I_1'$ respectively by V_0 and I_0 in the transfer matrix of the second quadrupole:

$$\begin{bmatrix} V_2 \\ I_2 \end{bmatrix} = [T_2]\begin{bmatrix} V_0 \\ I_0 \end{bmatrix} = [T_2].[T_1]\begin{bmatrix} V_1 \\ -I_1 \end{bmatrix}$$

Hence :

$$\begin{bmatrix} V_2 \\ I_2 \end{bmatrix} = [T_{Eq}]\begin{bmatrix} V_1 \\ -I_1 \end{bmatrix}$$

With :

$$[T_{Eq}] = [T_2].[T_1]$$

The transfer matrix of the equivalent quadrupole $[T_{Eq}]$ is equal to the product of the second transfer matrix and the first.

5. Examples of applications

5.1. Application example 1: Shaped quadrupole T

We consider the combination of three quadripoles in cascade as follows:

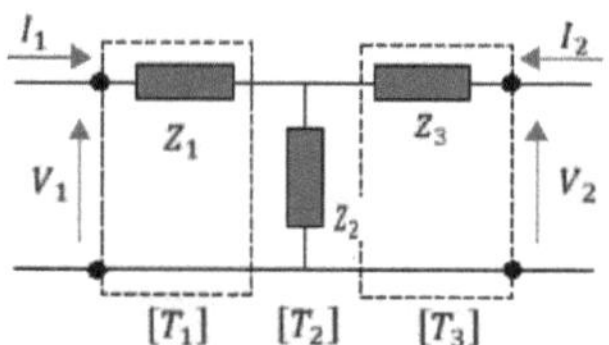

Determine the equivalent transfer matrix $\left[T_{Eq}\right]$ of the quadrupole in the form T.

According to the circuit of the electric quadrupole in the form T we have :

$$[T_1] = \begin{bmatrix} 1 & Z_1 \\ 0 & 1 \end{bmatrix}$$

$$[T_2] = \begin{bmatrix} 1 & 0 \\ 1/Z_2 & 1 \end{bmatrix}$$

$$[T_3] = \begin{bmatrix} 1 & Z_3 \\ 0 & 1 \end{bmatrix}$$

Gold :

$$\left[T_{Eq}\right] = [T_3].[T_2].[T_1]$$

Hence :

$$\left[T_{Eq}\right] = \begin{bmatrix} 1 + \dfrac{Z_3}{Z_2} & Z_1 + Z_3 + \dfrac{Z_1.Z_3}{Z_2} \\[2ex] 1/Z_2 & 1 + \dfrac{Z_1}{Z_2} \end{bmatrix}$$

5.2. Application example 2: Shaped quadrupole π

We consider the combination of three quadripoles in cascade as follows:

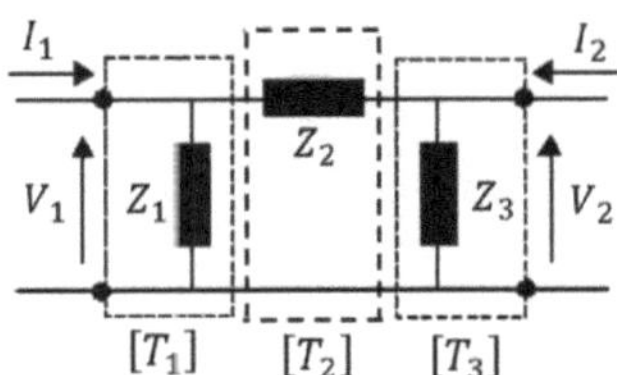

Determine the equivalent transfer matrix $\left[T_{Eq}\right]$ of the quadrupole in the form π.

According to the circuit of the electrical quadrupole in the form π above, we have :

$$[T_1] = \begin{bmatrix} 1 & 0 \\ 1/Z_1 & 1 \end{bmatrix}$$

$$[T_2] = \begin{bmatrix} 1 & Z_2 \\ 0 & 1 \end{bmatrix}$$

$$[T_3] = \begin{bmatrix} 1 & 0 \\ 1/Z_3 & 1 \end{bmatrix}$$

Gold :

$$[T_{Eq}] = [T_3].[T_2].[T_1]$$

Hence :

$$[T_{Eq}] = \begin{bmatrix} 1 + \dfrac{Z_2}{Z_1} & Z_2 \\ \dfrac{1}{Z_3} + \dfrac{1}{Z_1} + \dfrac{Z_2}{Z_1.Z_3} & 1 + \dfrac{Z_2}{Z_3} \end{bmatrix}$$

Chapter IV: Passive filters

1. Definition

Filters are widely used in electronics because their main purpose is to eliminate unwanted signals from signals.

In most cases, filters are made up of simple elements such as resistors, inductors and capacitors. Depending on the number and arrangement of the elements, different characteristics are obtained.

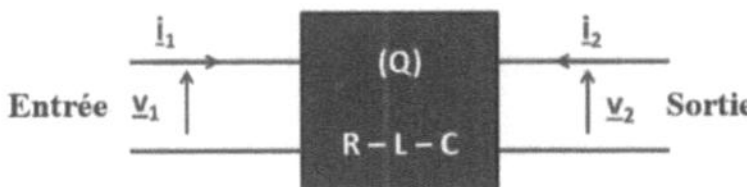

2. Filter characteristics

They are essentially based on the two Bode curves:

- the gain in decibels,

- the phase in degrees or radians.

However, it is very important to determine the cut-off frequency and the corresponding phase.

3. Complex transfer function

The complex transfer function is a particular characteristic of a quadripole inserted between a sinusoidal AC source and a load. In the case of a filter, it expresses the amplification in complex voltage.

The transfer function is defined by :

$$\underline{H}(j\omega) = \frac{\underline{s}}{\underline{e}}$$

Or :

$$\underline{H}(j\omega) = \left|\underline{H}\right| e^{j\varphi_{s/e}} = G e^{j\varphi_{s/e}}$$

With :

$\underline{e}$ Input voltage

$\underline{s}$ Output voltage

G Voltage gain

φ Phase shift of $\underline{e}$ in relation to $\underline{s}$

Therefore, we have :

$$\varphi_{rad}(\omega) = arg\left(\underline{H}(j\omega)\right) = arctg\left(\frac{Im\left(\underline{H}(j\omega)\right)}{Re\left(\underline{H}(j\omega)\right)}\right)$$

$$G_{dB}(\omega) = 20.\log_{10}\left(\left|\underline{H}(j\omega)\right|\right)$$

4. Different types of passive filters

Types of passive filters	Symbol
Low-pass filters	
High-pass filters	
Band-pass filters	
Notch filters or band rejection filters	

N.B.: Our study will be limited to first-order passive filters.

5. Study of a low-pass filter

5.1. Definition

The low-pass filter allows low frequencies to pass through and attenuates high frequencies.

5.2. Asymptotic behaviour of the circuit RC

Consider the following filter circuit:

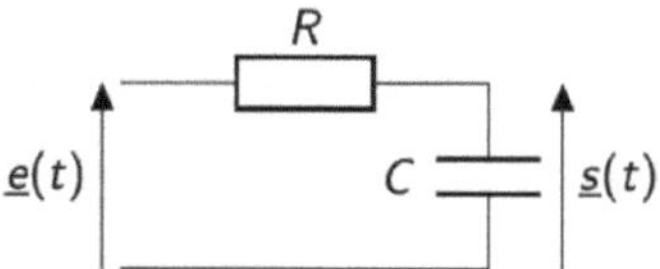

Low Frequency (LF) : $\omega \to 0$

We know that :

$$\underline{Z}_c = \frac{1}{jC\omega}$$

When $\omega \to 0$ then : $\underline{Z}_C \to \infty$. Then the capacitor is equivalent to an open switch (open circuit). In other words, :

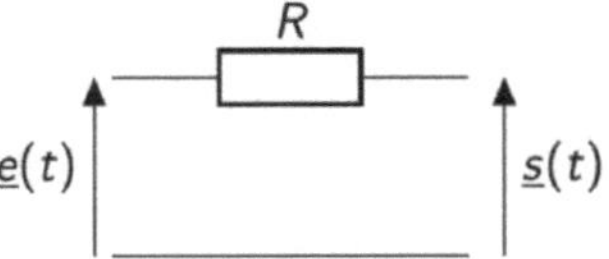

The new circuit diagram therefore becomes :

With : $s(t) = e(t)$

High Frequency (HF) : $\omega \to \infty$

When $\omega \to \infty$ then : $\underline{Z}_C \to 0$. Then the capacitor is equivalent to a closed switch (short circuit). In other words, :

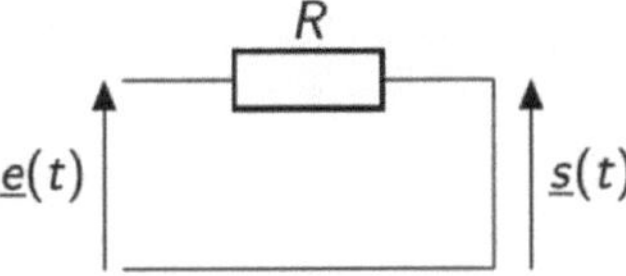

The new circuit diagram therefore becomes :

With : $s(t) = 0$

5.3. Filter transfer function

According to the voltage divider rule, we have :

$$\underline{H}(j\omega) = \frac{\underline{s}}{\underline{e}} = \frac{\underline{Z}_C}{\underline{Z}_C + \underline{Z}_R} = \frac{1}{1 + jRC\omega}$$

5.4. Mathematical determination of the cut-off frequency

By definition, we have :

$$\underline{H}(j\omega) = \frac{1}{1 + jRC\omega} \qquad \text{and} \quad G(\omega) = |\underline{H}(j\omega)| = \frac{1}{\sqrt{1 + (RC\omega)^2}}$$

Since :

$$G(\omega = \omega_c) = \frac{1}{\sqrt{2}}$$

So :

$$\frac{1}{\sqrt{1 + (RC\omega_c)^2}} = \frac{1}{\sqrt{2}}$$

This gives us :

$$\omega_c = \frac{1}{RC}$$

Hence :

$$f_c = \frac{1}{2\pi RC}$$

With : $\omega_c = 2\pi f_c$

5.5. Asymptotic study of gain and phase

We know that :

$$\underline{H}(j\omega) = \frac{1}{1 + jRC\omega}$$

The equation is :

$$\omega_c = \frac{1}{RC}$$

So :

$$\underline{H}(j\omega) = \frac{1}{1 + j\dfrac{\omega}{\omega_c}}$$

The expressions for gain and phase are :

$$G(\omega) = \frac{1}{\sqrt{1 + \left(\dfrac{\omega}{\omega_c}\right)^2}}$$

$$\varphi(\omega) = -arctg\left(\frac{\omega}{\omega_c}\right)$$

Therefore, we find :

Asymptotic study	$G(dB)$	$\varphi(rad)$
$\omega \to 0 \ (\omega \ll \omega_C)$	0	0
$\omega \to \infty \ (\omega \gg \omega_C)$	$-20.\log\left(\dfrac{\omega}{\omega_C}\right)$	$-\dfrac{\pi}{2}$
$\omega = \omega_C$	-3	$-\dfrac{\pi}{4}$

5.6. Gain Bode diagram

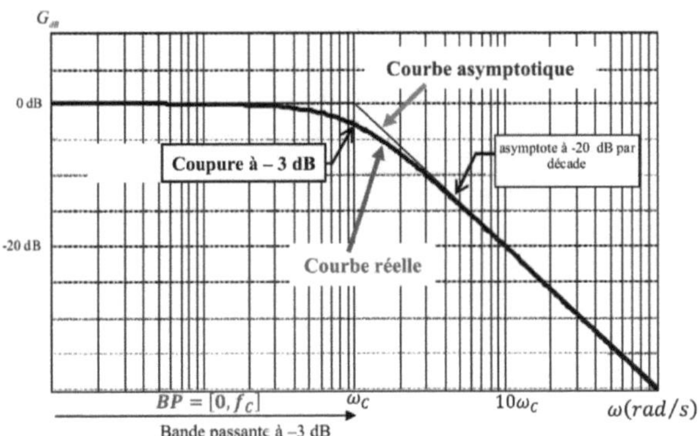

5.7. Bode diagram of the phase

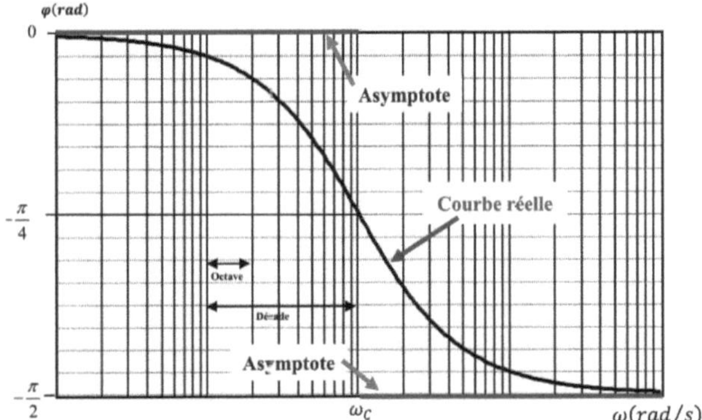

6. Study of a high-pass filter

6.1. Definition

The high-pass filter lets high frequencies through and attenuates low frequencies.

6.2. Asymptotic behaviour of the circuit RC

Consider the following filter circuit:

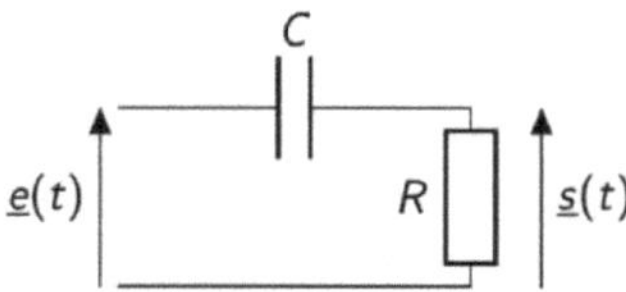

Low Frequency (LF) : $\omega \to 0$

We know that :

$$\underline{Z}_c = \frac{1}{jC\omega}$$

When $\omega \to 0$ then : $\underline{Z}_c \to \infty$. Then the capacitor is equivalent to an open switch (open circuit). In other words, :

Therefore : $s(t) = 0$

High Frequency (HF) : $\omega \to \infty$

When $\omega \to \infty$ then : $\underline{Z}_c \to 0$. Then the capacitor is equivalent to a closed switch (short circuit). In other words, :

Therefore : $s(t) = e(t)$

6.3. Filter transfer function

According to the voltage divider rule, we have :

$$\underline{H}(j\omega) = \frac{s}{e} = \frac{\underline{Z}_R}{\underline{Z}_R + \underline{Z}_c} = \frac{jRC\omega}{1 + jRC\omega}$$

6.4. Mathematical determination of the cut-off frequency

By definition, we have :

$$\underline{H}(j\omega) = \frac{jRC\omega}{1+jRC\omega} \qquad \text{and} \quad G(\omega) = \left|\underline{H}(j\omega)\right| = \frac{RC\omega}{\sqrt{1+(RC\omega)^2}}$$

Since :

$$G(\omega = \omega_c) = \frac{1}{\sqrt{2}}$$

So :

$$\frac{RC\omega_c}{\sqrt{1 + (RC\omega_c)^2}} = \frac{1}{\sqrt{2}}$$

This gives us :

$$\omega_c = \frac{1}{RC}$$

Hence :

$$f_c = \frac{1}{2\pi RC}$$

With : $\omega_c = 2\pi f_c$

6.5. Asymptotic study of gain and phase

We know that :

$$\underline{H}(j\omega) = \frac{jRC\omega}{1 + jRC\omega}$$

The equation is :

$$\omega_c = \frac{1}{RC}$$

So :

$$\underline{H}(j\omega) = \frac{j\dfrac{\omega}{\omega_c}}{1 + j\dfrac{\omega}{\omega_c}}$$

The expressions for gain and phase are :

$$G(\omega) = \frac{\dfrac{\omega}{\omega_c}}{\sqrt{1 + \left(\dfrac{\omega}{\omega_c}\right)^2}}$$

$$\varphi(\omega) = \frac{\pi}{2} - arctg\left(\frac{\omega}{\omega_c}\right)$$

Therefore, we find :

Asymptotic study	$G(dB)$	$\varphi(rad)$
$\omega \to 0 \ (\omega \ll \omega_C)$	$20.\log\left(\dfrac{\omega}{\omega_C}\right)$	$\dfrac{\pi}{2}$
$\omega \to \infty \ (\omega \gg \omega_C)$	0	0
$\omega = \omega_C$	-3	$\dfrac{\pi}{4}$

6.6. Gain Bode diagram

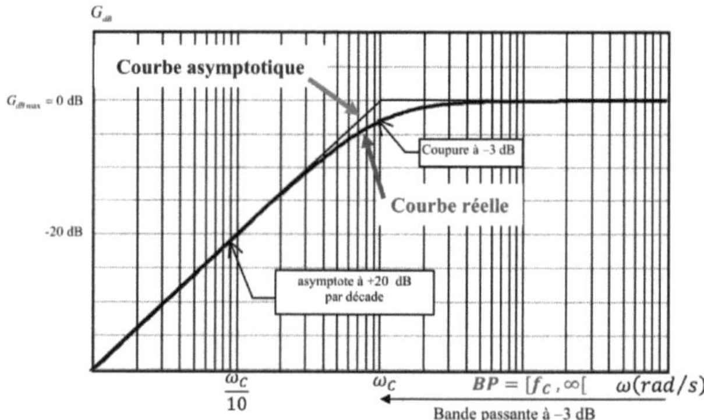

6.7. Bode diagram of the phase

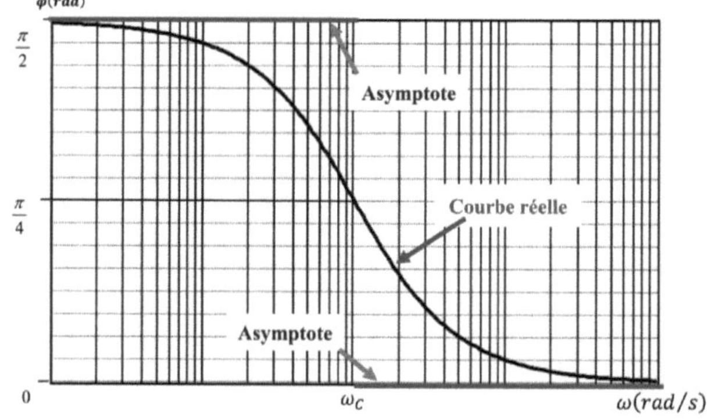

Chapter V: Diodes

1. Definition

A diode is any component with the property of being conductive in one direction of current and non-conductive in the other. The forward direction of the current is called direct and the other direction is called reverse.

2. Symbol and sign convention

2.1. Symbol

The diode is represented by the following standard symbol :

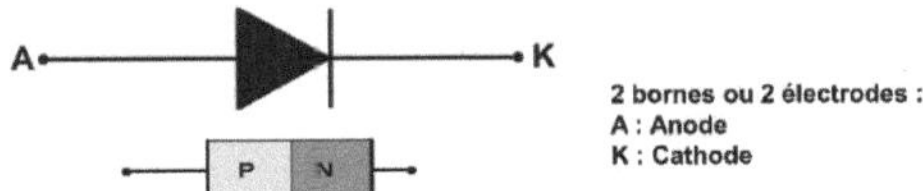

2.2. Sign convention

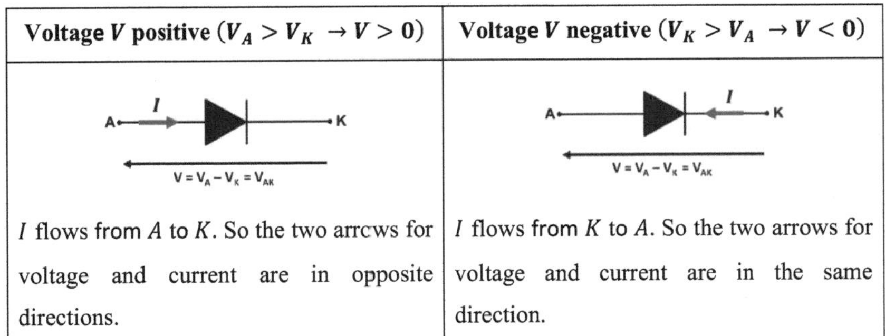

Voltage V positive $(V_A > V_K \rightarrow V > 0)$	Voltage V negative $(V_K > V_A \rightarrow V < 0)$
I flows from A to K. So the two arrows for voltage and current are in opposite directions.	I flows from K to A. So the two arrows for voltage and current are in the same direction.

3. Diode polarisation

3.1. Direct polarisation

In direct polarisation, the voltage applied V_{AK} $(V_{AK} > 0)$ allows an electric current to flow from the anode to the cathode, known as the direct current.

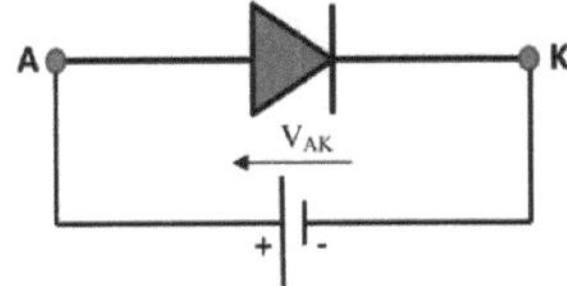

3.2. Reverse polarisation

In reverse polarisation, the voltage applied V_{AK} ($V_{AK} < 0$) prevents the current from flowing. The reverse current is therefore practically zero.

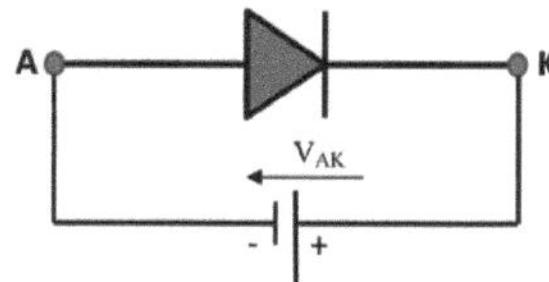

4. Current-voltage characteristic of a real diode

This characteristic describes the evolution of the current flowing through the diode as a function of the voltage at its DC terminals.

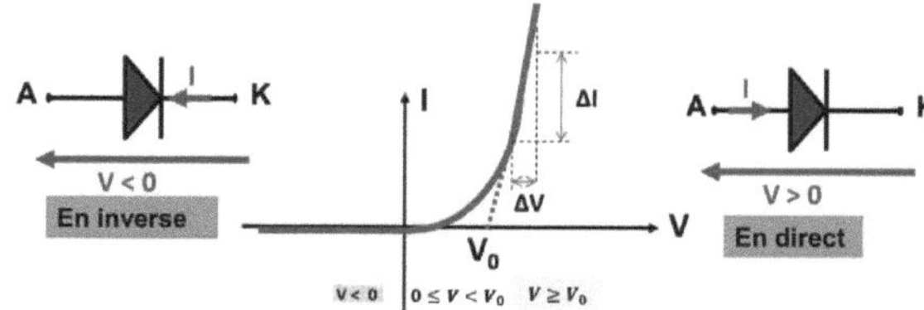

❖ *Direct route:*

$V > 0$ The diode is therefore positively or forward biased.

The diode is only conductive from $V \geq V_0$. It is also said to be conducting or on.

$0 \leq V < V_0$ The diode is directly blocked.

❖ *In the opposite direction:*

$V < 0$ The diode is therefore negatively or reverse biased.

The diode is non-conductive or non-transmissive. It is also said to be blocked or switched off.

5. Parameters deduced from the direct characteristic of a real diode

5.1. Practical voltage threshold

The voltage threshold of a real diode defines the practical voltage threshold V_0 which corresponds to the intersection between the straight part of the characteristic and the voltage axis.

Example:

- Silicon (Si) : $V_0 \approx 0.6\ V$ à $0.7\ V$.
- Germanium (Ge) : $V_0 \approx 0.3\ V$ à $0.4\ V$.

5.2. Dynamic resistance

The dynamic resistance of a diode is defined by :

$$R_d = \frac{\Delta V}{\Delta I}$$

5.3. Static resistance

This is a resistor connected in place of a diode through which the same DC current flows I_d and has the same DC voltage V_d.

$$R_s = \frac{V_d}{I_d}$$

6. Equivalent diagrams of a real diode

6.1. Direct polarisation

In direct polarisation, the diode is equivalent to a receiver with electromotive force V_0 and internal dynamic resistance R_d.

6.2. Reverse polarisation

In reverse bias, the diode is equivalent to an open switch.

7. Electrical equation for the diode

The electrical equation of a diode is defined by :

$$I = I_s \left(e^{\left(\frac{e.V}{\eta.K.T}\right)} - 1 \right)$$

With :

I_s Saturation current

K Boltzmann constant

T Temperature in Kelvin

e Elementary charge of the electron

$1 < \eta < 2$: Ideality factor

The equation is :

$$V_T = \frac{K.T}{e}$$

So :

$$I = I_s\left(e^{\left(\frac{V}{\eta.V_T}\right)} - 1\right)$$

8. Operating point of a diode

Consider the simple diode circuit below. The diode is supplied with a DC voltage E through a resistor R.

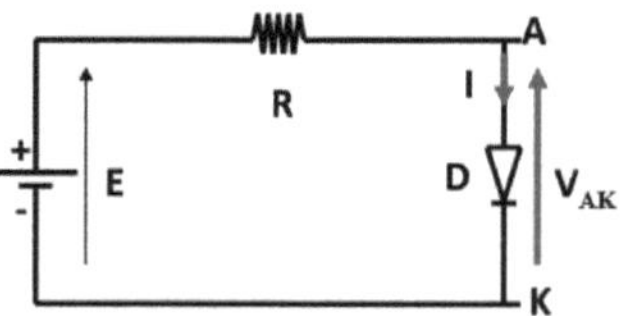

The voltage V_{AK} applied across the diode is given by :

$$V_{AK} = E - R.I$$

We can also write :

$$I = \frac{E - V_{AK}}{R}$$

This equation is that of the load line.

Plot the characteristic of the load line $I = f(V_{AK})$.

For $I = 0\ mA$ we have : $V_{AK} = E$

For $V_{AK} = 0\ V$ we have : $I = \frac{E}{R}$

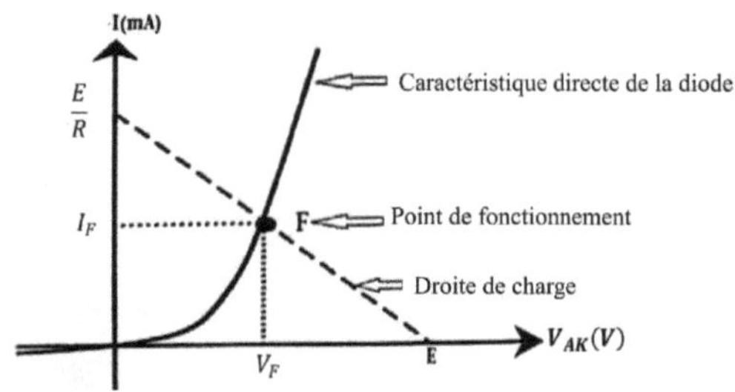

The point F where the load line intersects the diode's direct characteristic is called the operating point, the bias point or the rest point.

9. Zener diode

9.1. Definition

The Zener diode is designed to have, in its reverse characteristic, a zone of "controlled breakdown or non-destructive avalanche", in which the reverse current increases rapidly without any significant increase in voltage and without damaging the diode.

9.2. Symbol

The Zener diode is represented by the following standard symbol :

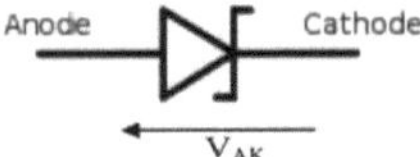

9.3. Current-voltage characteristic of a Zener diode

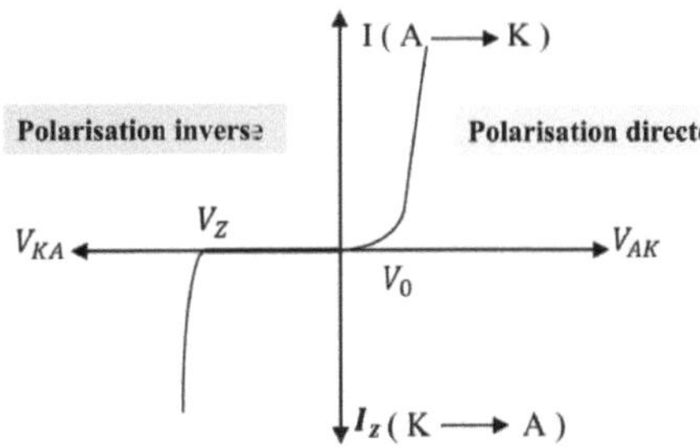

In direct polarisation, the Zener diode is equivalent to a real diode.

In reverse bias, the Zener diode conducts when the reverse voltage V_{KA} becomes greater than the Zener voltage V_Z.

The linearised characteristic leads to the equation : $V_{AK} = V_Z + R_Z . I_Z$ where R_Z is the inverse dynamic resistance. In this case, the Zener diode is equivalent to the following model :

Chapter VI: Bipolar transistors

1. Definition

A bipolar transistor is a semiconductor-based electronic device from the transistor family. Its operating principle is based on two PN junctions, one forward and one reverse.

Polarising the reverse PN junction with a small electric current (sometimes called the transistor effect) enables a much larger current to be controlled using the principle of current amplification.

2. Description

The bipolar transistor is a device with three layers, of different types, separated by two junctions. The three layers are called emitter, base and collector. There are two types of transistor.

Transistor NPN   **Transistor PNP**

The emitter is a heavily doped area. It is the source of electrons (or holes) and is responsible for emitting charges.

The base is a thin, lightly doped zone. It is very thin, in the order of a micrometre. It is the path for the charges coming from the emitter.

The collector is a larger zone than the other medium-doped zones. It collects the charges coming from the emitter.

3. Symbol and sign convention

The transistor is a voltage mesh and a current node.

Law of meshes : $V_{BE} = V_{BC} + V_{CE}$

Law of knots : $I_E = I_B + I_C$

4. Transistor current gain

The current gain of the transistor is defined by :

$$\beta = \frac{I_C}{I_B}$$

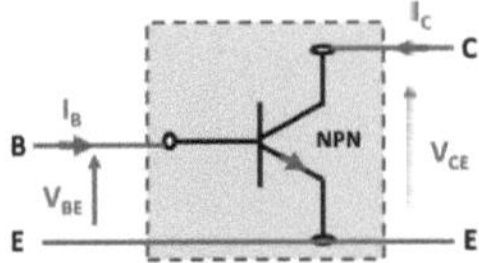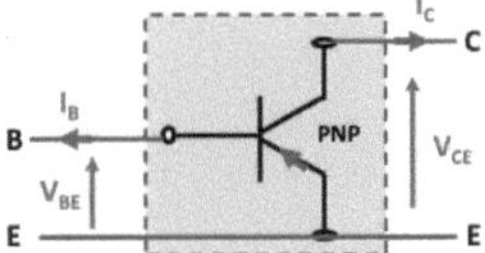

β varies between 10 à 800 depending on the type of transistor.

5. Relationship between α and β of the transistor

α is a proportionality coefficient linking the collector current to the emitter current. It is given by :

$$I_C = \alpha . I_E$$

We know that :

$$I_E = I_B + I_C$$

Or :

$$\frac{I_C}{\alpha} = \frac{I_C}{\beta} + I_C$$

Hence :

$$\alpha = \frac{\beta}{1+\beta} \quad \text{and} \quad \beta = \frac{\alpha}{1-\alpha}$$

From now on, for the bipolar transistor, we consider that :

$$I_C = \beta . I_B \quad \text{and} \quad I_C \approx I_E \quad \text{with} : \alpha \approx 1$$

Indeed, I_B and I_C are the only two currents flowing in the bipolar transistor.

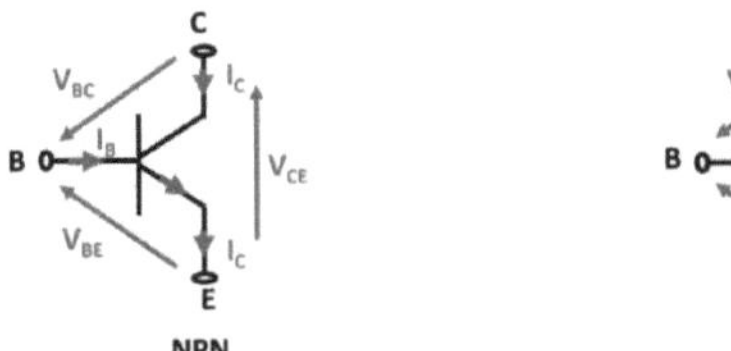

N.B.: In the following, our study will be limited to NPN bipolar transistors.

6. Different operating modes of the bipolar transistor

There are generally three operating modes for NPN bipolar transistors: FC mode, FB mode and FS mode.

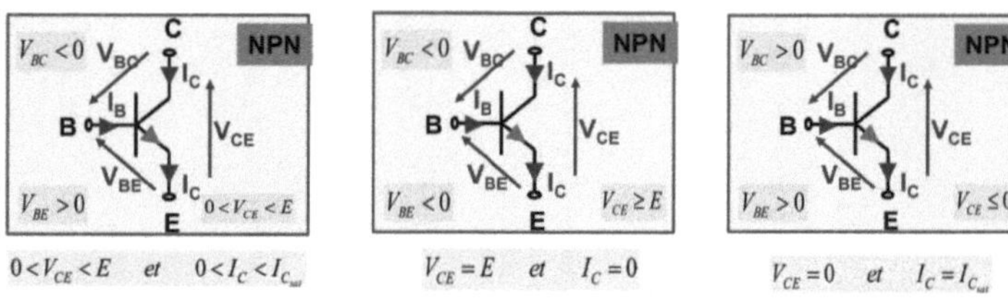

(a): FC mode (b): FB mode (c): FS mode

With :

FC: Conductor Operation or Direct Normal Operation.

FB: Operation Locked or Off.

FS: Saturated or On.

7. Bipolar transistor polarisation

7.1. Principle and benefits

Biasing is the operation that consists of choosing an operating point for the transistor. This operation is essential if the transistor is to operate correctly in a circuit.

To bias a transistor, you need to choose a bias circuit. In general, there are several types of circuit, the most commonly used of which is the base bridge bias circuit.

7.2. Polarisation assembly using a base bridge

A single power supply is used E to supply both the base circuit and the collector circuit.

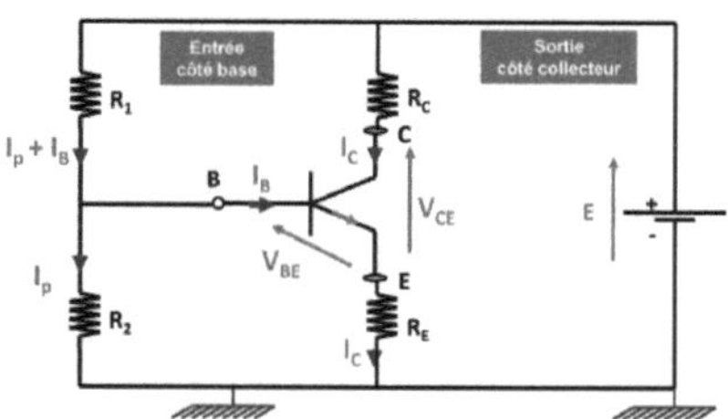

Resistance R_1 and R_2 form the base bridge.

R_C, R_E, R_1 and R_2 are the polarisation resistors.

I_p bridge current.

It is assumed that :

$$I_p \gg I_B$$

This implies :

$$I_p + I_B \approx I_p$$

❖ *Useful polarisation equations :*

Entry (base side) :

Maille (1) : $E \approx (R_1 + R_2).I_p$

Maille (2) : $R_2.I_p \approx V_{BE} + R_E.I_C$

Output (collector side) :

Maille (3) : $E \approx (R_C + R_E).I_C + V_{CE}$

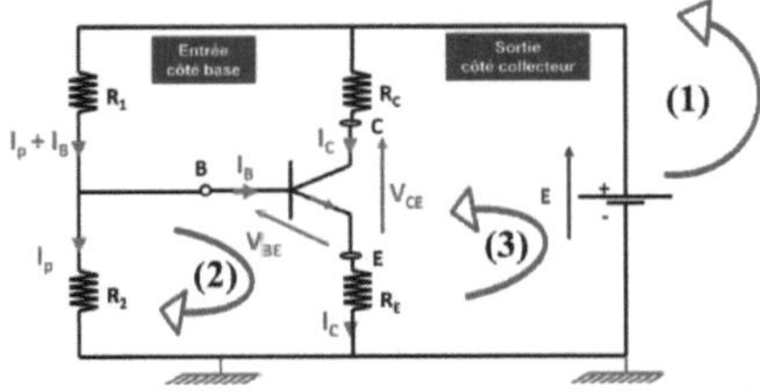

7.3. Static load line

The purpose of the load line is to provide the location of all the transistor's operating points.

According to mesh (3), we have : $E \approx (R_C + R_E).I_C + V_{CE}$

The equation for the static load line is therefore :

$$I_C = \frac{E - V_{CE}}{(R_C + R_E)}$$

We plot the characteristic of the static load line $I_C = f(V_{CE})$

For $V_{CE} = 0\ V$ then : $I_C = \frac{E}{(R_C + R_E)}$

For $I_C = 0\ mA$ then : $V_{CE} = E$

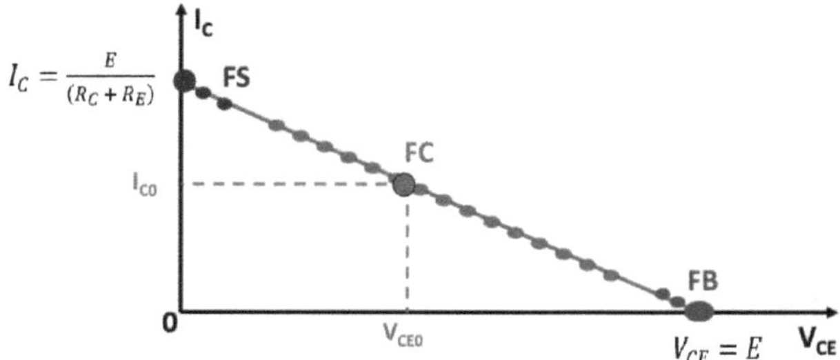

The output torque (V_{CE0} et I_{C0}) defines the desired operating point.

The input torque (V_{BE0} et I_{B0}) will be obtained by deduction or by the attack line.

7.4. Static attack line

The purpose of the drive line is to provide the location of all the transistor's operating points.

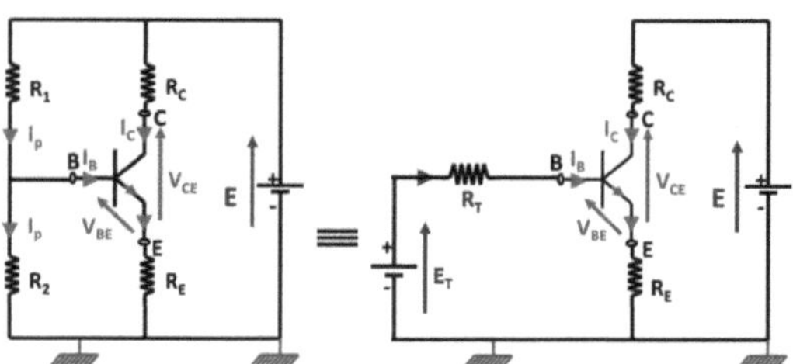

We apply the principle of Thévenin's theorem and obtain :

$$E_T \approx E . \frac{R_2}{(R_1 + R_2)} \quad \text{and} \quad R_T = \frac{R_1 . R_2}{(R_1 + R_2)}$$

$$E_T \approx R_T . I_B + V_{BE} + R_E . I_C = (R_T + \beta . R_E). I_B + V_{BE}$$

Then the equation of the static attack line is written :

$$I_B = \frac{E_T - V_{BE}}{(R_T + \beta . R_E)}$$

We plot the characteristic of the static attack line $I_B = f(V_{BE})$

For $V_{BE} = 0\ V$ then : $I_B = \dfrac{E_T}{(R_T + \beta . R_E)}$

For $I_B = 0\ mA$ then : $V_{BE} = E_T = E . \dfrac{R_2}{(R_1 + R_2)}$

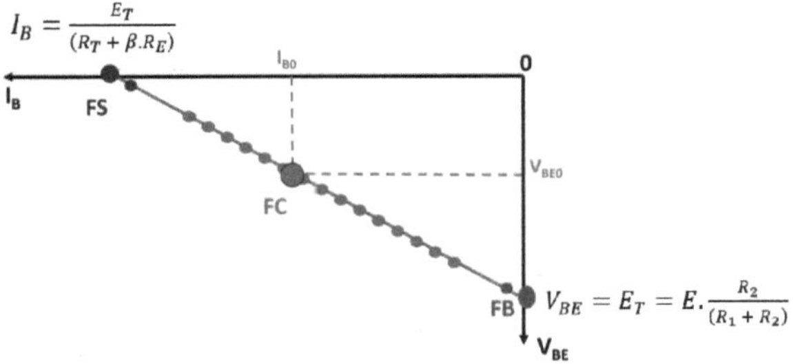

The input torque (V_{BE0} et I_{B0}) defines the operating point we want to reach.

Finally, the quadruplet (V_{CE0}, I_{C0}, V_{BE0} et I_{B0}) defines the operating point of the NPN bipolar transistor, as shown in the diagram below.

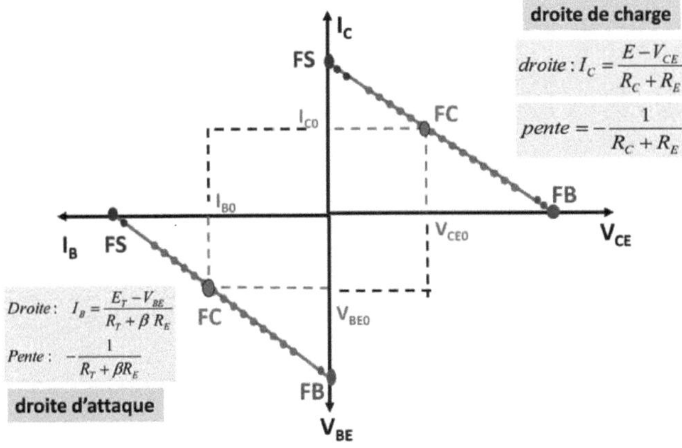

Exercises

Exercise 1:

Calculate the equivalent resistance between points A and B of the following two circuits (a) and (b) :

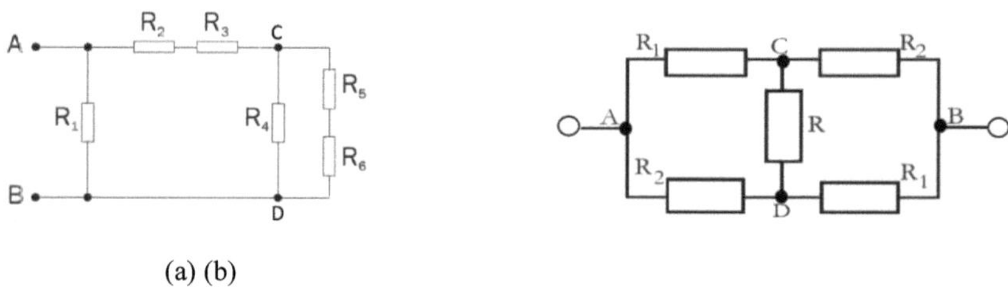

(a) (b)

Data	
Circuit (a)	**Circuit (b)**
$R_1 = 220\ \Omega$, $R_2 = 560\ \Omega$, $R_3 = 3.9\ K\Omega$, $R_4 = 25.6\ K\Omega$, $R_5 = 220\ K\Omega$ and $R_6 = 470\ K\Omega$	$R = 100\ \Omega$, $R_1 = 2R$ and $R_2 = R$

Exercise 2:

Calculate the equivalent resistance between points A and B of the following circuits (a), (b) and (c):

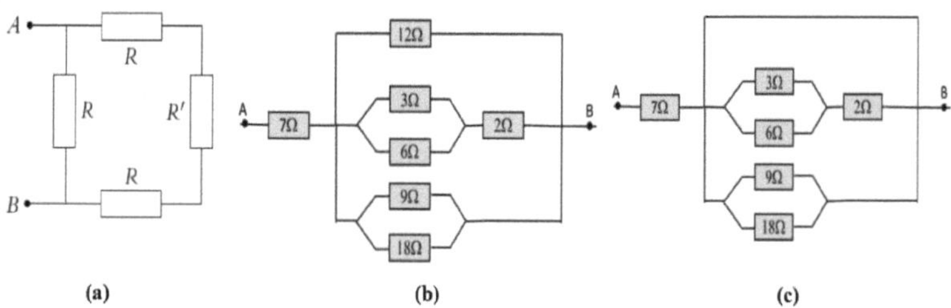

(a)   (b)   (c)

Exercise 3:

On each of the three diagrams (a), (b) and (c) below. Calculate the value of the unknown quantity.

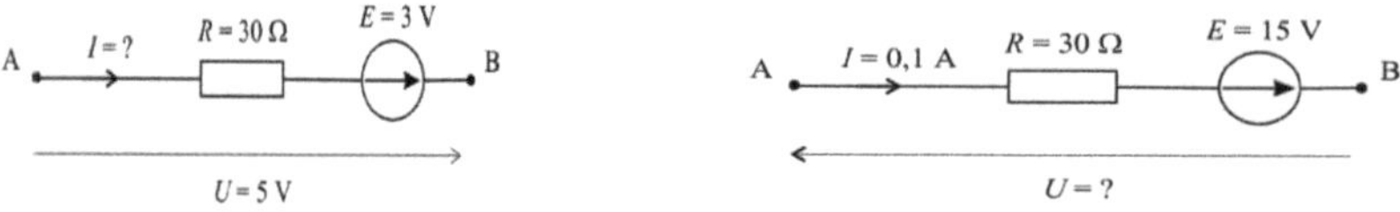

(a) (b)

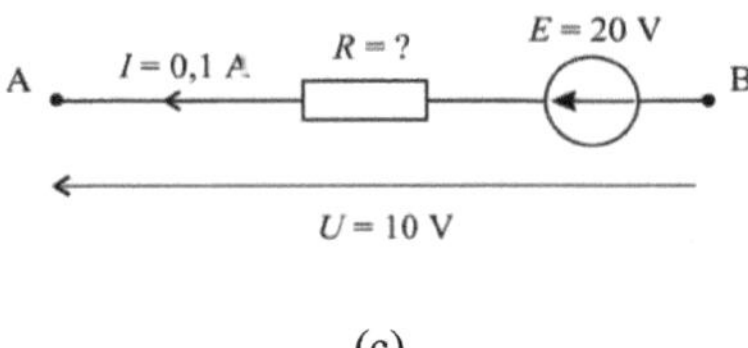

(c)

Exercise 4:

Consider the following electrical circuit:

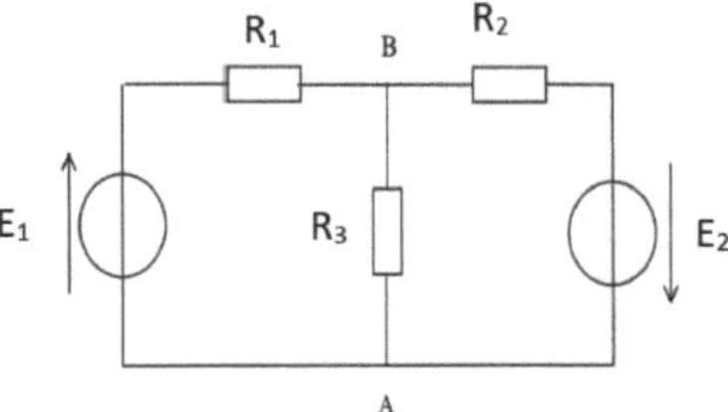

Data : $R_1 = 16\,\Omega$, $R_2 = 4\,\Omega$, $R_3 = 6\,\Omega$, $E_1 = 4\,V$ and $E_2 = 24\,V$.

Calculate the current in the AB branch by applying :

1. Kirchhoff's laws.
2. Millman's theorem.
3. The superposition theorem.

Exercise 5:

For the circuit below, give the current flowing in the BC branch, $i_3 = 2\,mA$, $R = 5\,K\Omega$ and $R_1 = 2R$.

1. Calculate the voltages u_5, u_3 and u_4.
2. Calculate the currents i_4, i_1 and i_5.
3. Deduct the current i and voltage u.

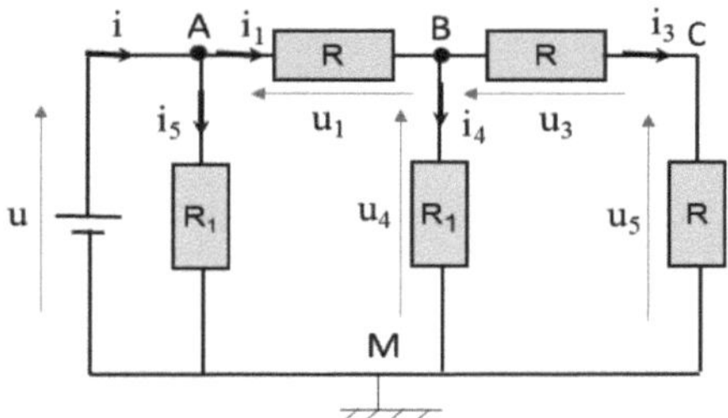

Exercise 6:

Given the circuit from exercise 2, to $u = 40\ V$ find :

1. The value of the current i using the series-parallel combination of resistors.
2. The value of the current i_1 using the series-parallel association of resistors and Thévenin's theorem.
3. The value of i_3 using the triangle-star transformation and Thevenin's theorem.

Exercise 7:

We want to determine the current flowing through the resistor R_3 in the circuit below using several methods:

1. By applying successive transformations between Thévenin and Norton generators.
2. Using the superposition theorem.
3. Given $R_3 = 0\ \Omega$ calculate the voltage across R_4.
4. Given $R_1 = R_3 = 0\ \Omega$ calculate the current delivered by the generator E_5.

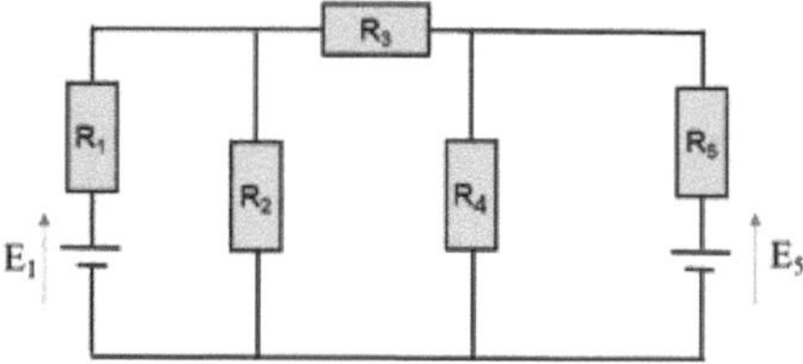

Exercise 8:

Consider the following electrical circuit:

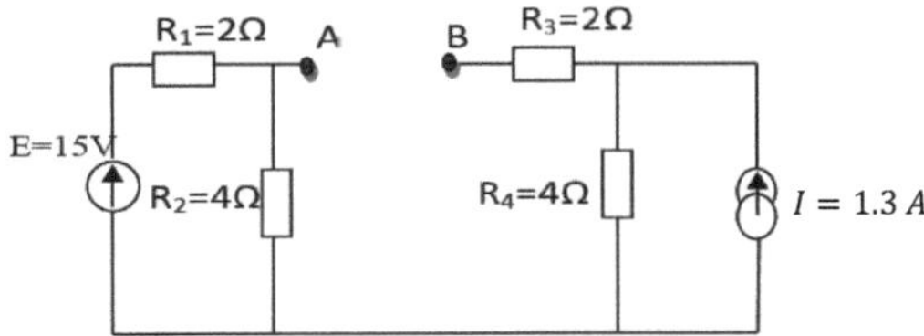

1. Determine the equivalent Thévenin model (E_{Th} et R_{Th}) between points A and B.
2. We connect a resistor $R_5 = 10\ \Omega$ between points A and B. Calculate the current flowing through this resistor.

Exercise 9:

Calculate the current I in the resistor R of the circuit shown below, by applying successive transformations between the Thévenin and Norton generators.

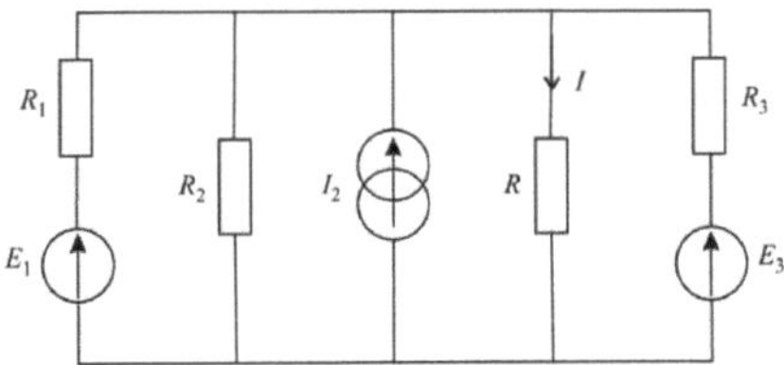

Data : $E_1 = 10\ V$, $I_2 = 100\ mA$, $E_3 = 7\ V$, $R_1 = 60\ \Omega$, $R_2 = 100\ \Omega$, $R_3 = 40\ \Omega$ and $R = 30\ \Omega$.

Exercise 10:

Consider the electrical circuit below. A sinusoidal voltage is applied between points A and B $e(t) = E_m \cos(\omega t)$.

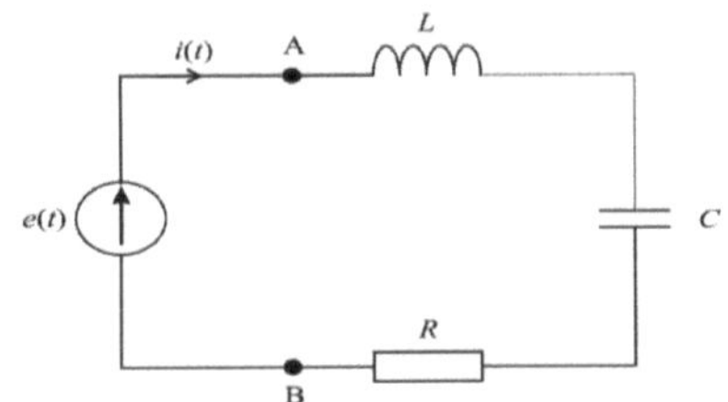

1. Find the expression for the complex impedance of the dipole.
2. Calculate the modulus of the complex impedance of the dipole. A.N. : $R = 10\ K\Omega$, $C = 1\ \mu F$, $L = 2\ mH$ and $f = 50\ Hz$.
3. For what value ω_0 of ω is the current i have its maximum amplitude? For this same value, what is the value of the phase shift between this current and the voltage? $e(t)$?

Exercise 11:

Consider the circuit shown in the diagram below. The AB branch contains an inductance in parallel (coefficient L and internal resistance r) and a capacitor of capacitance C. A sinusoidal voltage : $u(t) = U_m \cos(\omega t)$.

1. Calculate the current (maximum or rms I) of the current i.
2. Determine the value ω_0 of the pulsation ω for which I is minimal.
3. For $\omega = \omega_0$ give the expression for I. Examine the case $r = 0$.

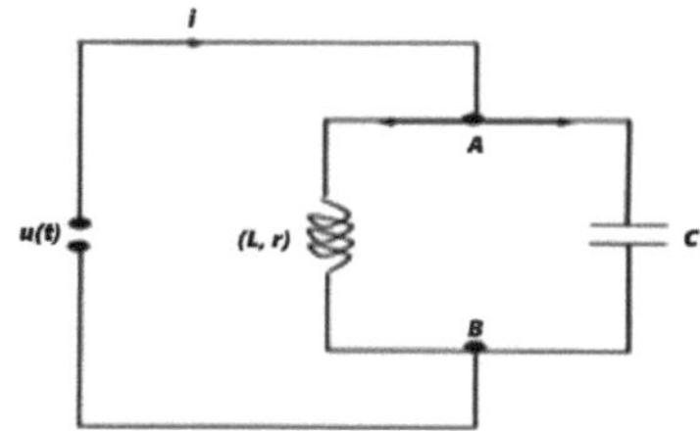

Exercise 12 :

Consider the circuit below, supplied with an AC voltage source : $u(t) = U_m \cos(\omega t)$.

1. Give the expression for the complex impedance Z_0 of the circuit as a function of L, C, R and ω.
2. Determine the value L_0 of the inductance for which the circuit impedance is equivalent to a pure resistance.
3. For $L = L_0$ determine the real impedance Z_{r0} of the circuit and the maximum current I_m.
4. The resistor R is replaced by an inductance L' and ω is assumed to be variable.
   a. Give the expression for the complex impedance Z of the new circuit.
   b. Calculate the phase shift φ.

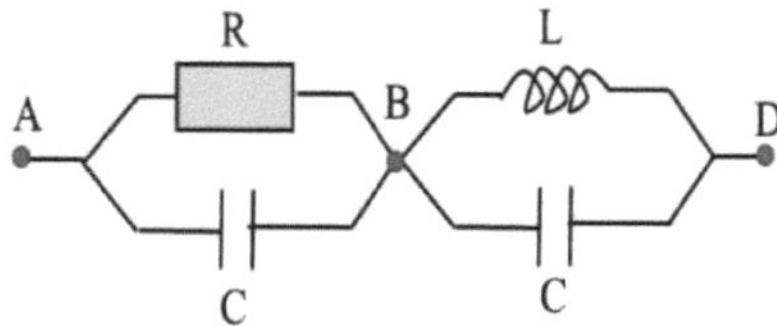

Exercise 13:

Consider the following active quadrupole:

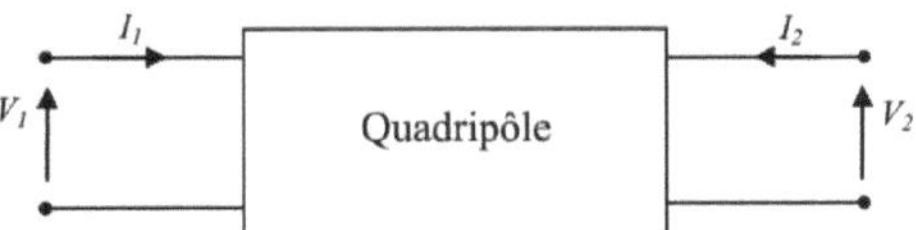

1. Define the hybrid parameters (H) of the quadrupole.

2. Determine the admittance parameters (Y) of the quadrupole as a function of the parameters (H).

3. Conversely, determine the parameters (H) of the quadrupole as a function of the parameters (Y).

Exercise 14:

Consider the following quadrupole:

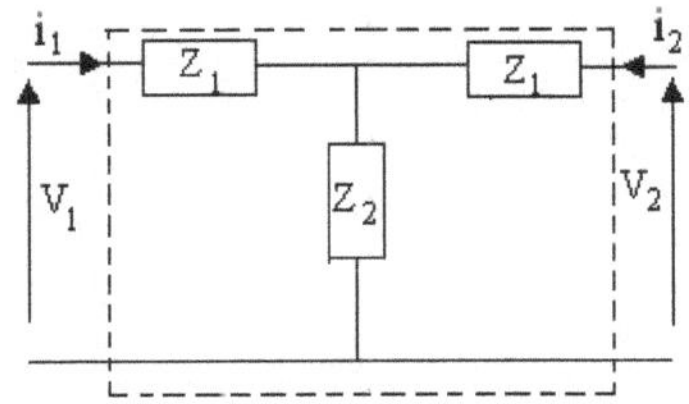

1. Determine the hybrid parameters H_{ij} of this quadrupole from their definitions.

2. Derive the impedance parameters Z_{ij}.

3. The quadrupole is supplied with a sinusoidal voltage and loaded with an impedance of Z_0. Calculate the input impedance Z_e of the circuit as a function of Z_{ij} and Z_0.

4. Z_1 is a capacity of value $2C$ and Z_2 is an inductance of value L. We want Z_e be equal to Z_0. Determine the expression for Z_0 as a function of L, C and ω.

Exercise 15:

Consider the following electrical quadrupole:

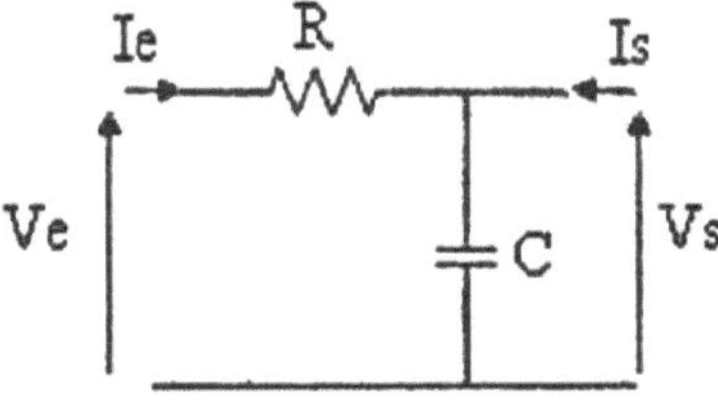

1. Determine the elements of the impedance matrix $[Z]$ and the corresponding electrical model.

2. Determine the elements of the admittance matrix $[Y]$ and the corresponding electrical model.

3. Determine the elements of the hybrid matrix $[h]$ and the corresponding electrical model.

4. Determine the elements of the equivalent transfer matrix $\left[T_{\acute{e}q}\right]$ of the quadrupole.

Exercise 16:

Consider the following circuit:

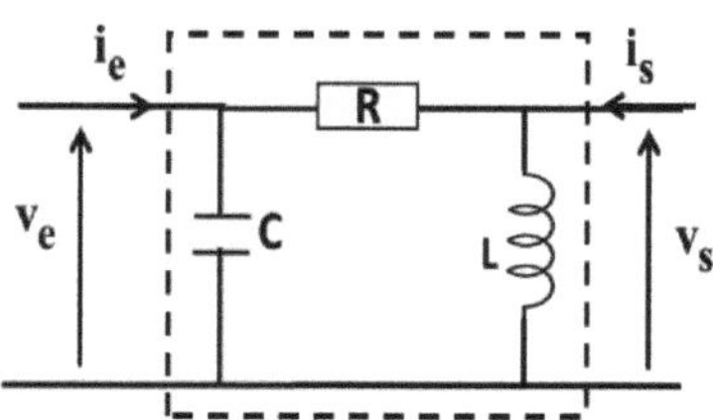

1. Calculate the representative transfer matrix $[T]$ of the above quadrupole. Assume that the circuit is used as an open circuit. ($i_s = 0$).

2. Establish the expression of the transfer function $H(j\omega)$ of the circuit in the form :

$$H(jx) = \frac{jx}{1 + jx}$$ with x the reduced pulsation. Determine the expression of x as a function of R, L and ω.

3. Establish the expression for gain $G(x)$ and $G_{dB}(x)$.

4. Establish the expression for phase $\varphi(x)$ in the form : $\varphi(x) = \frac{\pi}{2} - arctan(x)$

5. Determine the expression for the asymptote of $G_{dB}(x)$ and $\varphi(x)$ if $x \longrightarrow 0 \ \varepsilon\tau \ x \longrightarrow \infty$ $\square\square$.

6. Draw the Bode diagram.

7. Determine the nature of the filter.

Exercise 17:

Consider the following filter:

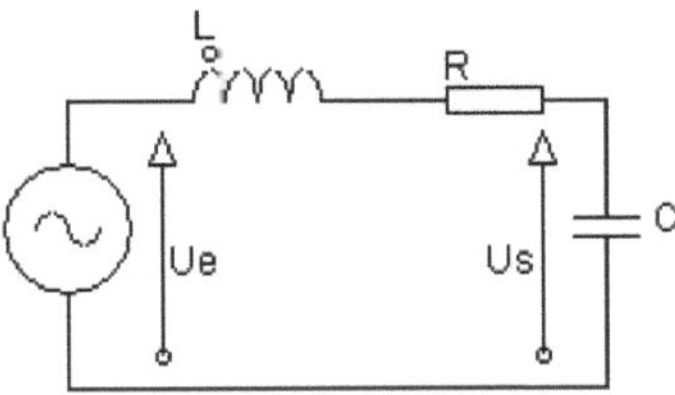

The gain curve $G_{dB} = 20 \log\left(\frac{U_s}{U_e}\right)$ as a function of frequency is shown below.

1. Determine graphically the cut-off frequency at $-3\ dB$ of the filter.
2. Determine the values of the gain G_{dB} in the case where $f < 10\ Hz$ and in the case where $f = 20\ KHz$.
3. Calculate the amplitude of the output voltage if the input voltage has amplitude $24.8\ V$ and frequency $f = 20\ KHz$.
4. If the input voltage is a DC voltage v then what is the output voltage?

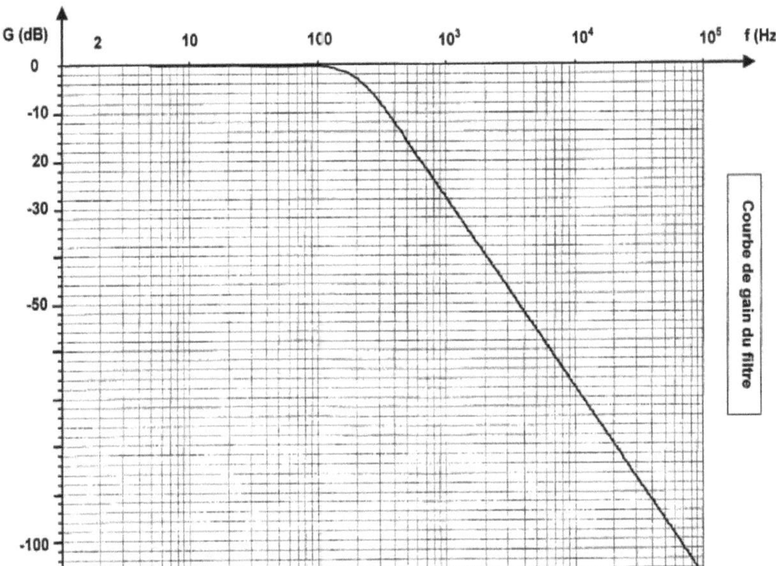

Exercise 18:

Consider the following electrical circuit:

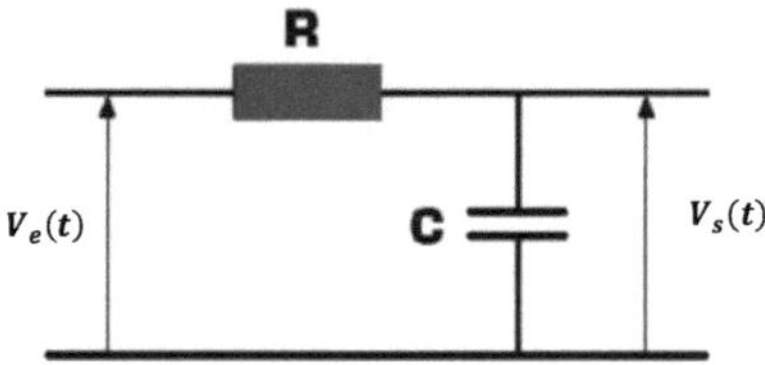

Data : $V_e = 10\ V$, $R = 1\ K\Omega$ and $C = 20\ nF$.

1. Establish the asymptotic behaviour of the circuit RC in the following two cases:

   a. $\omega \longrightarrow 0$

   b. $\omega \longrightarrow \infty$

2. Express the transfer function $\underline{H}(j\omega) = \dfrac{V_s(t)}{V_e(t)}$ as a function of R and C.

3. What is the order and type of filter?

4. Give the expression and value of the cut-off frequency f_c.

5. Calculate the output voltage V_s if the frequency of the input voltage is equal to $\dfrac{f_c}{10}$ or $10f_c$.

6. Plot the Bode diagrams for gain and phase. ☐

Exercise 19:

Consider the following electrical circuit:

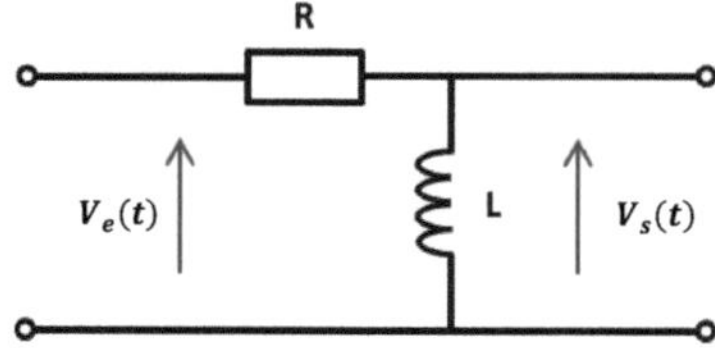

Data : $V_e = 10\ V$, $R = 1\ K\Omega$ and $L = 0.2\ H$.

1. Express the transfer function $\underline{H}(j\omega) = \dfrac{V_s(t)}{V_e(t)}$ as a function of R and L.

2. What is the order and type of filter?

3. Give the expression and value of the cut-off frequency f_c.

4. Calculate the output voltage V_s if the frequency of the input voltage is equal to $\dfrac{f_c}{10}$ or $10f_c$.

5. Plot the Bode diagrams for gain and phase.

Exercise 20:

Consider the following electrical filter RC filter:

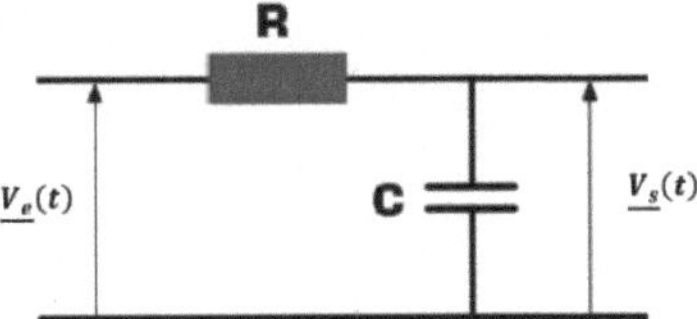

Data : $V_e(t) = V_{e\,max} \sin(2\pi ft)$, $V_s(t) = V_{s\,max} \sin(2\pi ft + \varphi)$ and $C = 0.47\ \mu F$.

1. Find the expression for the filter transfer function as a function of R, C and f.

2. Deduce that the gain of this filter is written as : $G_{dB} = -10 log[1 + (2\pi RCf)^2]$

3. The curve below shows the change in filter gain G of the filter as a function of frequency f.

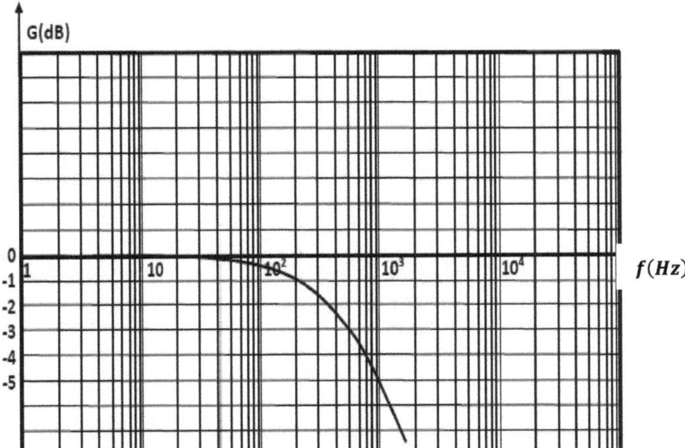

Determine graphically the *maximum* value G_0 of the *gain G* the *cut-off* frequency f_c and the bandwidth.

4. A *sinusoidal* voltage is applied to the input of this filter $V_e(t) = 9 \sin(800\pi t)$.

a. Justify whether or not this voltage will be transmitted. Determine the nature of the filter.

b. If so, calculate the *maximum* value $V_{s\,max}$ of the voltage transmitted?

5. Establish the expression for the *cut-off* frequency f_c of this filter as a function *of* R *and* C.

6. Calculate the value of the *resistance* R.

Exercise 21:

Consider the following electronic circuit:

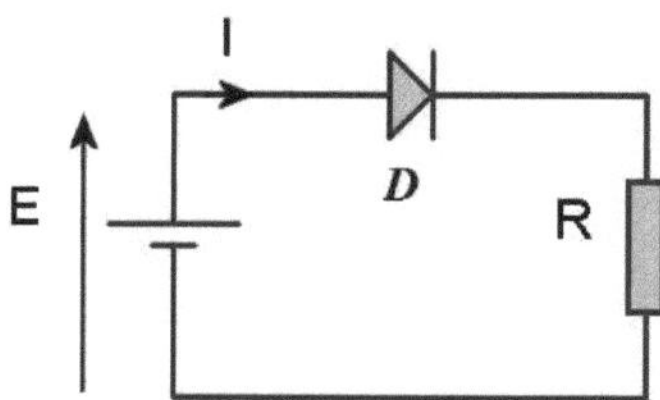

1. Plot the voltages V_D and V_R.
2. Knowing that $V_{Dseuil} = 0.7\ V$. What is the value of E if the diode is blocked?
3. For $E = -5\ V$ and $R = 1\ K\Omega$. Calculate the value of I.
4. For $E = 1\ V$. Calculate the values of I and V_R.

Exercise 22:

1. Write the equation for the diode charging line used in Figure 1.
2. Take $E = 1.4\ V$ and plot the load line in the plane (I, V) in Figure 2.
3. Determine the coordinates of the saturation point, the blocking point and the operating point.
4. Deduct the static resistance of the diode.
5. Using its characteristic, determine the elements of the equivalent diagram of the diode in the forward and reverse directions.

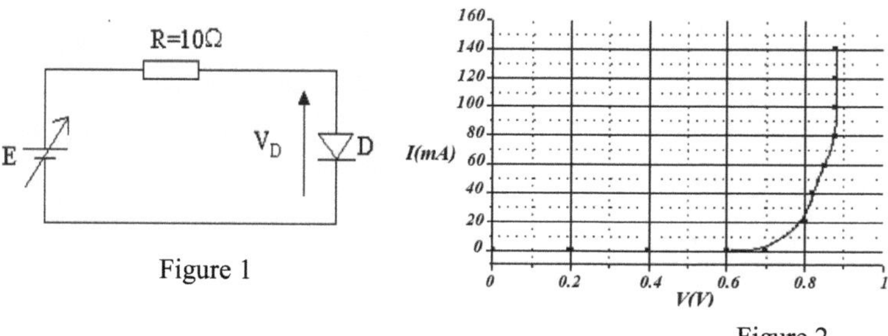

Figure 1

Figure 2

6. The diode D is used in the circuit shown in Figure 3. Give the condition for the diode to be conductive.

$$\text{We give}: V_e(t) = 20\sin(\omega t) \text{ and } E = 5\ V.$$

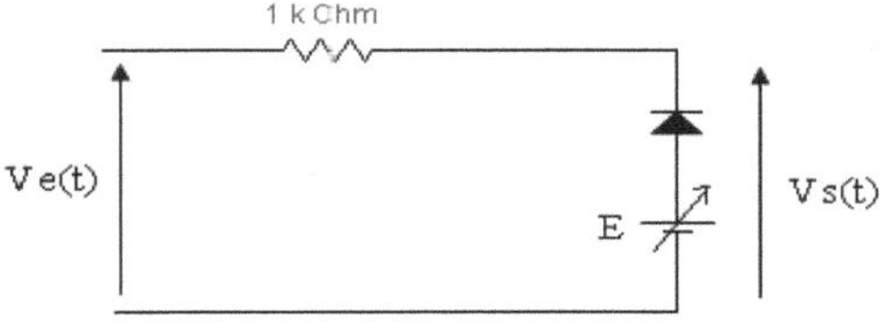

Figure 3

Exercise 23 :

A voltage Zener diode $V_z = 45\ V$ is used to regulate a rectified sinusoidal voltage, which can vary between the limits $40 < V_e < 60\ V$

The dynamic resistance of the diode is assumed to be zero.

1. When $V_e = 40\ V$ is measured $I_s = 20\ mA$. Deduce the value of R_s.
2. At what value of V_e is the output voltage regulated?
3. Draw the transfer graph $V_s = f(V_e)$.
4. Calculate the current I_z in the diode when $V_e = 60\ V$.

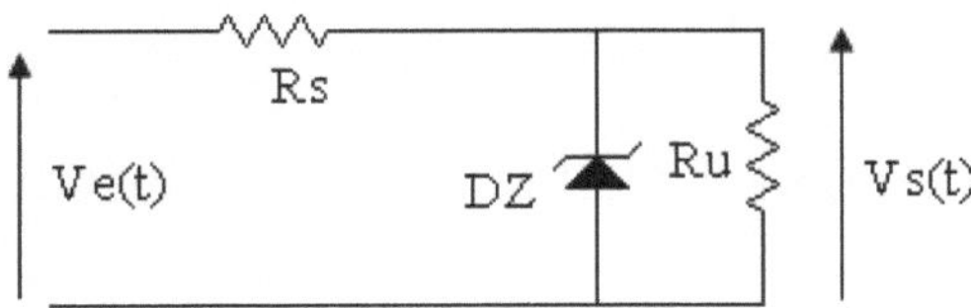

We give : $R_u = 1.8\ K\Omega$

Exercise 24 :

A Zener diode with voltage $V_z = 6.2\ V$ and zero dynamic resistance is inserted in the circuit below. When reverse biased, the Zener diode can withstand a maximum current of $I_{z\,max} = 200\ mA$.

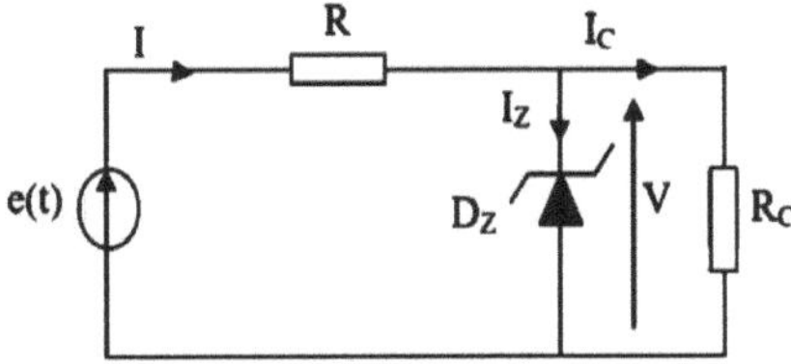

Data : $R = 100\ \Omega$ and $R_C = 220\ \Omega$.

1. Calculate the value of the current I_C.

63

2. What are the values of I_{min} and I_{max} to ensure regulation without destroying the Zener diode?

3. Deduct the minimum value e_{min} and the maximum value e_{max} of the voltage $e(t)$.

4. Disconnect the resistor from the load R_C and set the voltage $e(t)$ to the value e_{max}.

4.1. Calculate the value of I_Z.

4.2. Calculate the new value of R to protect the Zener diode.

Exercise 25 :

Consider the following electronic circuit:

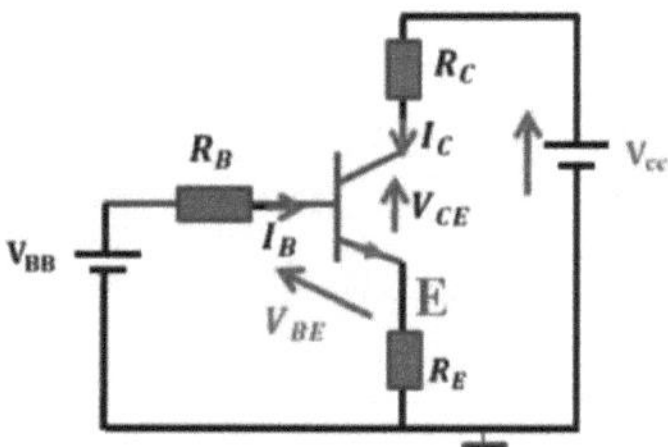

Data : $V_{BB} = 4\ V$, $V_{cc} = 12\ V$, $\beta = 100$, $V_{BE} = 0.7\ V$, $R_B = 40\ K\Omega$, $R_C = 1\ K\Omega$ and $R_E = 1\ K\Omega$.

Calculate I_B , I_C , I_E and V_{CE}.

Exercise 26 :

Consider the following electronic circuit:

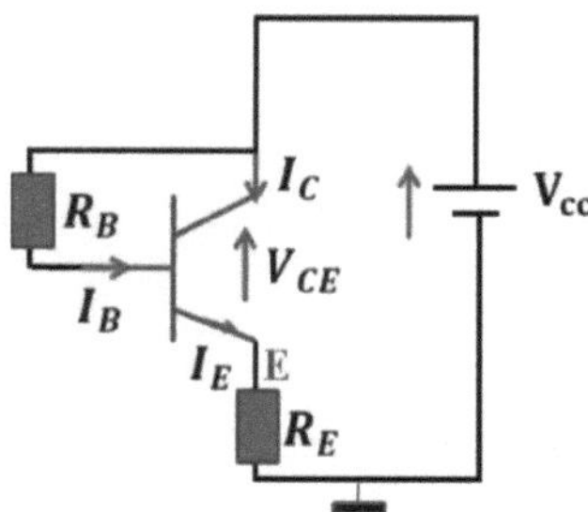

Data : $V_{cc} = 12\ V$, $\beta = 150$, $V_{BE} = 0.6\ V$, $R_B = 100\ K\Omega$ and $R_E = 750\ \Omega$.

Determine the operating point (I_C et V_{CE}) of the transistor.

Exercise 27:

Consider the following electronic circuit:

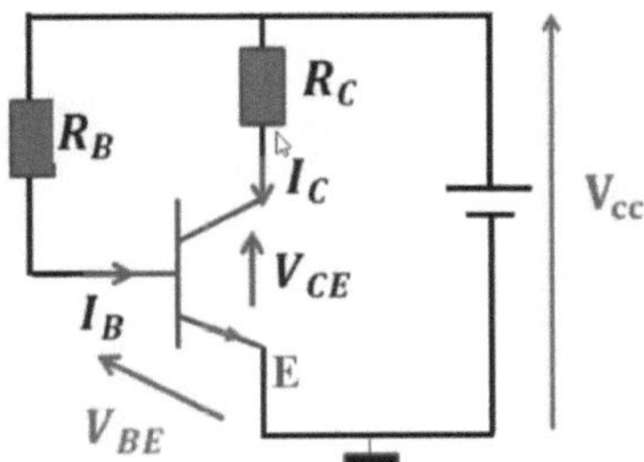

Data : $V_{cc} = 12\ V, \beta = 100, V_{BE} = 0.7\ V$ and $R_C = 60\ \Omega$.

1. It is assumed that the current through the resistor R_C is $100\ mA$. Calculate the value of the resistor R_B.

2. If we vary the base resistance R_B the base current I_B varies and therefore I_C also changes. What is the value of the current I_C when the NPN transistor is saturated?

3. What is the value of R_B to saturate the transistor?

Exercise 28:

Consider the circuit shown in Figure 1, where the bias is provided by the resistor between the collector and the base. The transistor is characterised by the network of curves in Figure 2. In continuous operation, we have : $\beta = 65$.

$$\text{We give} : U_0 = 10\ V, R_B = 17\ K\Omega, R_C = 1\ K\Omega \text{ and } R_E = 100\ \Omega$$

1. Give the equation of the static load line at the output.
2. Give the equation of the static input drive line. Deduce the rest point of the circuit.
3. Plot the output static load line on the network of curves in Figure 2.

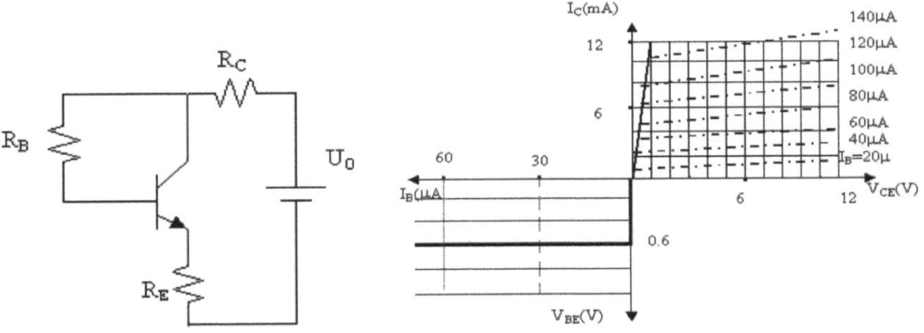

Figure 1 Figure 2

Exercise 29:

Consider the assembly below for which :

$E = 20\ V, \beta = 200, V_{BE} = 0.7\ V, R_C = 1\ K\Omega$ and $R_B = 100\ K\Omega$.

Calculate I_B, I_C and V_{CE}.

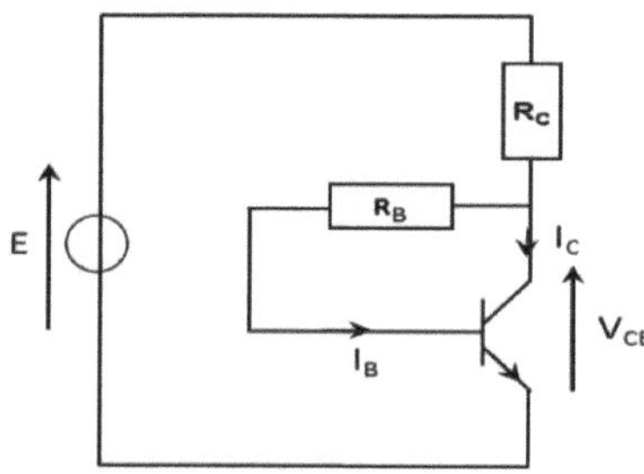

Exercise 30:

Consider the assembly below for which :

$I_1 = 175\ \mu A, I_2 = 5.05\ mA, \beta = 200, V_{BE} = 0.3\ V, R_1 = 2\ K\Omega$ and $R_2 = 100\ K\Omega$.

1. Show the symbol for the transistor on the copy, indicating the voltages V_{BE}, V_{CE} and drawing a diagram of the BE and BC junctions.
2. When $I_1 = 0\ A$ what is the state of the transistor? Deduce the values of I_B and I_C.
3. Assuming: Transistor operates in Normal (active) mode.

3.1. Calculate I_B and I_C.

3.2. Check the operating mode of the transistor selected. If this is not the case, recalculate I_C.

Bibliography

[1]   O. El Ouatassi, electronics course, Errachidia Faculty of Science and Technology, A.U.: 2004-2005.

[2]   A. Baghdad, electrical and electronic circuits course, Mohammedia Faculty of Science and Technology, A.U.: 2018-2019

[3]   A. El Alami, course in electrical and electronic circuits, Errachidia Faculty of Science and Technology, A.U.: 2020-2021

[4]   A. Siad, general electronics course, Institut Supérieur des Etudes Technologies de Nabeul, A.U. : 2013-2014.

[5]   Y. El Hassouani, course in electrical and electronic circuits, Errachidia Faculty of Science and Technology, A.U.: 2020-2021.

[6]   S. D. Bennani, course in analogue filtering, Ecole Nationale des Sciences Appliquées de Fès, A.U.: 2019-2020.

[7]   JN. Beury, filter course, Lycée Grand Lebrun, Bordeaux.

[8]   T. Lamhamdi, course in electrical and electronic circuits, Faculty of Science and Technology of Fez, A.U.: 2016-2017.

[9]   M. Habibi, electronics course, Kénitra Faculty of Science, A.U.: 2004-2005.

[10]   A. El Alami & Y. El Hassouani, tutorials for the "electrical and electronic circuits" module, Errachidia Faculty of Science and Technology, A.U.: 2020-2021.

[11]   A. El Alami & M. Hamdaoui, tutorials for the "electrical and electronic circuits" module, Errachidia Faculty of Science and Technology, A.U.: 2022-2023.

Printed by Books on Demand GmbH, Norderstedt / Germany